LF

# MORE CUNNING THAN MAN

The legendary Pied Piper of Hamelin *(Engraving by Paul Thumann, courtesy Museum Hameln, Hameln, West Germany)*

Robert Hendrickson

STEIN AND DAY / *Publishers* / New York

First published in 1983

Designed by Judith Dalzell
Printed in the United States of America
STEIN AND DAY/*Publishers*
Scarborough House
Briarcliff Manor, N.Y. 10510

**Library of Congress Cataloging in Publication Data**

Hendrickson, Robert, 1933–
More cunning than man.

Bibliography: p.
Includes index.
1. Rats—Social aspects. 2. Rats—Control—History.
3. Rats as carriers of disease—History. 4. Rodents—
History. 5. Social history. I. Title.
RA641.R2H46 1983 614.4′38 82-48512
ISBN 0-8128-2894-1

For all he means to me
this book is dedicated
with love
to my son, Erik

Great rats, small rats, lean rats, brawny rats,
Brown rats, black rats, gray rats, tawny rats . . .

They fought the dogs and killed the cats
    And bit the babies in the cradles,
And ate the cheeses out of the vats
    And licked the soup from the cooks' own ladles . . .

—Robert Browning, *The Pied Piper of Hamelin,* 1845

# CONTENTS

# ACKNOWLEDGMENTS

I would like to thank the many people who provided expert information, patiently explained technical points, and supplied illustrations for this work, particularly: John J. Gallagher, Chief, Program Services Branch, Environmental Health Service Division, Center for Environmental Health of the U.S. Public Health Service; Madison L. Brown III, First Assistant Commissioner Chicago Department of Streets and Sanitation; Terence Howard, Acting Assistant Director Chicago Rodent Control; A. Ros, Assistant County Environmental Health Director II, Dade County, Florida, Department of Public Health; Dr. Stephen C. Frantz, Rodent and Bat Specialist, Laboratories for Veterinary Science, Center for Laboratories and Research State of New York Department of Health; Dr. Al Rizzo, Director, Community Education, Bureau for Pest Control, New York City Department of Health; the U.S. Department of the Interior; the U.S. Department of Agriculture; and Robert Schwadron, Lederle Laboratories.

Special thanks are of course due my editor, Benton Arnovitz, and, as always, my wife and co-worker, Marilyn.

R. H.

# I

# OF RATS AND MEN

Scurrying along their runways, instinctively trying to escape the catastrophe they sensed, the rats of Engebi Island slipped their sleek supple bodies deep into their burrows even before the first blinding light, the deafening explosions, mushroom clouds, and gaping craters deforming the earth. Above them, throughout Eniwetok, it seemed as if the world was ending. Mammoth nuclear bombs were being dropped on this western Pacific atoll. Engebi shuddered again and again, apparently scoured of all life, plant and animal, as the United States tested its atomic arsenal. Scientists wagered that nothing living there could have survived.

When biologists inspected the island several years later, in 1950, most of their worst suspicions seemed confirmed. The remaining land and marine animals, all plant life, and even the soil itself, were found to be highly radioactive. Yet further investigation revealed that the rats of Engebi had not only survived—they had thrived. "The island abounded with rats," recalls Dr. William B. Jackson, a task force leader and now director of the Environmental Studies Center at Bowling Green University. "Not maimed or genetically deformed creatures, but robust rodents so in tune with their environment that their life spans were longer than average. The rats' burrows shielded some from direct effects of the blasts, but any way you look at it, their survival was uncanny."

Perhaps one day, if the penultimate bomb is ever dropped, the rats of this world will be as safe, secure, and serene in their elaborate and ingeniously intricate burrows as the rats of Engebi. In any event, such incredible tales, and many more to follow, of a species seemingly programmed for survival have led some observers to suggest that the world's wily rats, which at five billion strong already outnumber people, may well outlast or overpower man and inherit the earth.

Rats could quickly fill any environmental gap left by the destruction of human life on earth. The cunning rodents already have the widest range of every New World mammal except man, have adapted to all climates, and extended their territory steadily throughout history, proving that they can live with or without people. Despite their much smaller, less convoluted cerebrums, rats are in many ways more highly developed than man. Their sense of taste, for example, is so acute that they can detect as little as two parts per million of a poison hidden in a food. Unlike many overweight people, they seem always able to balance their diets, whether feeding on garbage or out of a lion's feeding pan at the zoo, and they disdain foods like the refined flour that we eat in huge quantities, though we know it is bad for us.

The omnivorous rat, whose very name means "gnawing animal," has teeth so strong and sharp that it can cut through thick concrete and steel sheeting. Rats constantly gnaw to stay alive and are even able to use the echoes of their grinding teeth as a kind of sonar to enable them to choose obstacle-free paths. So highly developed is their sense of balance that they can scale a sheer brick wall, walk tightropes, and land on their feet after falling sixty feet from a building. They are worriers like people, often biting *their* nails, too, and they are instinctive planners or hoarders, frequently collecting the same things human misers do, including money. There is indeed at least one story of rats collecting live crabs for food—biting off their legs and corralling them like cattle in their burrows for a future food supply. At times, however, their voracious appetites overwhelm their hoarding instincts. When Sheik Shakhbut, the oil-rich former leader of Abu Dhabi, took an inventory of the immense amounts of money he had hoarded in his chambers under the bed, stuffed inside his mattress, and stored in closets and boxes, he found that rats had eaten over $2 million of it.

Rats are intelligent creatures that can be taught to perform many tasks. Recently, Cambodian premier Lon Nól accused Khmer Rouge insurgents of training rats to deliver explosives during the Indochina war. Curiosity is the key to rat intelligence. Like people, they seem happiest when busy doing something. They are always exploring; experiments have in fact shown that they can be taught to do complicated tricks when their only reward is the chance to explore new territory. As for their sex life, there is ample proof that male and female rats are even busier and more promiscuous than their human counterparts.

Where they are not soiled by civilization, all rats—and especially the black rat—are beautiful immaculate animals. Unlike people, rats are always personally clean creatures who are continually grooming themselves and each other; they are transformed into "dirty rats" only when they come in contact with man's environment. A Wisconsin woman wondered why a fresh basin of water she left in her cellar at night was always dirty the next morning—until late one night she surprised a pair of rats bathing in it.

Rats can also be more compassionate than man. Helpless rats are often fed all their lives by others, and there have been many documented stories of rats biting off the tails of brother rats caught in traps, or guiding blind rats. The rodents probably cooperate in ways that man can still only imagine.

For centuries rats have been thought to follow a wise "King Rat," a huge intelligent leader paler and more cunning than his followers, who consider him monarch of the species. Swiss zoologist Konrad von Gesner makes the earliest mention of such a creature in his *Historia animalium* (1551-58), where he wrote: "Some say that the rat, in its old age, grows enormously large and is fed by the younger rats; it is called a King Rat by our people." Although there is no scientific proof of the King Rat's existence, sewer workers in London and other large cities have told of sighting the fabulous beast, who is usually attended by a bodyguard of huge white rats and causes his subject rats to grow silent and motionless when he appears in their midst. One much repeated tale has the subject rats stealing red cloth and fashioning him a king's robe out of it.

Assemblies of rats in a conclave around their cunning king for some purpose are a common theme of folklore and fairytales, as is shown by the famous Daumier drawing reproduced here (see frontpiece) and other storybook illustrations. But journals as soberly conservative and devoted to facts as the *Wall Street Journal* have reported sightings of crafty King Rats and their subjects. On August 25, 1969, a *Journal* correspondent wrote of a shower of rats falling on the Indonesian island of Lombok. "Rat showers" like this one have been cited before, but the *Journal* article goes well beyond them in its account of a mysterious, cunning King Rat and his army threatening the island's rice crop:

> On the outskirts of Batudjai, a half dozen farmers . . . not nearly as superstitious as one might suppose and who know their local animal life very well indeed . . . are squatting in a ricefield. . . . "The rats came six months ago, before the rains stopped!" says a farmer. How did they come? "They fell from the sky." From the sky? "Yes, in bunches of seven, and then they spread out across the land" the farmer adds matter-of-factly. "They are led by a great white rat as large as a cat," says another farmer. "The white rat is very smart. It knows when we plan to harvest. If we plan to harvest a field the day after tomorrow the rats will eat the field tomorrow night. If we plan, in secret, to harvest the field tomorrow then the rats will eat it tonight." A visit to the

> village chief, the only fat man to be seen in Batudjai ("He is of a higher caste," explains a villager) repeats the farmers' story. Led by a white "King of rats as large as a dog," the rats appeared last December, falling from the sky in bunches of seven, he says. As they landed the rats separated and spread in seven different directions. . . . Some farmers saw this happen, says the chief, and several farmers nearby nod.

King rats, real or unreal, should not be confused with the amazing but quite real "rat kings" discussed further on. These rat kings, composed of up to 32 rats living with their tails inextricably knotted together, manage to survive for long periods of time. No one knows certainly how they become attached, but in the early nineteenth century respected naturalists of the day suggested that younger, healthier rats tied up the tails of older weaker rats to provide good nests for themselves. The hypothesis may be fanciful, but it does show a great respect for the rat's intelligence.

In recent times, Professor Erik Tolman demonstrated his high regard for the rat by dedicating his scholarly book *Purposive Behavior in Animals and Men* to the creature. Nobel Prize-winning novelist William Faulkner believed that rats were the leaders among animals in intelligence, which he defined as "the ability to cope with environment: which means to accept environment yet still retain something of personal liberty." In *The Reivers* Faulkner wrote: "The rat of course I rate first. He lives in your house without helping you to buy it or build it or keep the taxes paid; he eats what you eat without helping you raise it or buy it or even haul it into the house; you cannot get rid of him; were he not a cannibal, he would have long since inherited the earth."

Countless laboratory tests, including complicated mazes, have shown that the rat is an extremely intelligent animal blessed with a good memory, excellent insight, and the ability to solve problems. Investigators report that rats are so well organized that both the brown and black species send out a scout or "pioneer" rat to stand on its hind legs at the burrow entrance and test the wind for several minutes to see if the coast is clear before the remaining rats will emerge.

To escape floods, rats sometimes crawl headlong into long narrow passages they have dug within their burrows, one rat to a passage; each rat fits snuggly into its niche, making it airtight, and can remain alive as long as the oxygen lasts in the considerable area beyond its head.

In sewers, workers have noticed rats using the little floats on which rat poisons are suspended as life rafts during rain storms, or as means of transportation when they tire of swimming.

Rats are so cunningly suspicious of food that they are extremely difficult to poison. The first rat to find an unknown food usually decides whether the rest of the pack will partake of it and he often leaves it untouched for almost a week. "If a few animals of the pack pass the food without eating any," a researcher notes, "no other pack member will eat any either. For if the first rats do not eat poisoned bait, they sprinkle it with their urine or feces." Another scientist believes that "knowledge of the danger of a certain bait is transmitted from rat generation to generation and the knowledge long outlives those individuals that first had the experience." It is certain that in early youth the composition of its mother's milk teaches a young rat about what the mother has eaten, helping it to avoid unsafe foods.

quote

In his epoch-making *On Aggression,* Konrad Lorenz argues that "the difficulty of effectively combating the most successful biological opposite to man, the Brown Rat, lies chiefly in the fact that the rat operates basically with the same methods as those of man, *by traditional transmission of experiences* and its dissemination within the close community." Rats of course also resemble man in their great numbers; these vast armies constitute their main menace, not the fantastically evil portraits of individual monster rats that some writers have drawn, although rats contaminated by man's environment can be as repulsive and dangerous in their own way as their most horrible representatives in fiction.

Like men and unlike most other animals, rats will sometimes, if rarely, display homosexual behavior when members of the opposite sex are readily available. Far more important is the fact that rats, like humans, are not highly specialized animals, or are specialized in being nonspecialized, as Lorenz put it, and can adapt well to many circumstances.

Men have long compared rats to humans and drawn parallels between the two species, with no regard for the possibility that the rat might find such comparisons odious, considering the record of our race. In story and real life, rats are often endowed with human characteristics. Once they were tried in courts of law for their offenses and a Russian czar reportedly court martialled a rat for its predations. Many are the tales of prisoners taming rats and making them companions during long solitary confinements. An American prisoner of war in Vietnam developed a humanlike relationship with a rat over a period of several years. He was overwhelmed with grief when the rat suddenly disappeared, but his companion returned just as mysteriously some time later, missing one leg—which she may have lost in a manmade trap—yet still completely trusting her human friend.

Even today wild rats are befriended, worshipped, and housed, fed and cared for by various peoples around the world. Some people value them for no apparent reason—like the Mississippi legislator who introduced an unsuccessful bill to make the wharf rat the official state animal back in the early 1970s. Usually the reasons are rooted inextricably in history or legend. In a modern Mayan Indian creation story, for example, God is tied up and left to die in a burning field by the jealous Ancient Men, the brothers of his mother, the Virgin Mother. God calls to the rats in the earth, who emerge from their burrows, gnaw the ropes loose (shades of Poe's "The Pit and the Pendulum"!) and take God with them deep into the earth beneath the flaming fields.

Yet the rat owes the fascination, and the religious awe in which it is held, not to its intelligence, compassion, or even to the fact that hundreds of millions of rats have died often horrible deaths for man in scientific experiments over the years. The rat is strangely, darkly fascinating—the most horrific creature of our nightmares—because it is almost universally considered an evil, guileful, mysterious creature whose deadly destructive actions are often unfathomable. Why do rats sense impending disaster and desert a ship or building long before it sinks or burns? How can rats know a marketplace will be moved many miles away and leave the old site for the new long before any human move is made? Do rats perpetually war against each other as they did in the terrible war between brown and black rats centuries ago, which has been called the greatest interspecific conflict of all time? Why do rats make seemingly unprovoked attacks on humans?

No one really knows the answers to such questions, even though the rat has been the subject of more scientific study than any other animal known to man over the last century. What is certain is that any history of the rat is a history of human misery. More numerous than man on earth, sustained by human food, living largely by human sloth, the dread black plagues that rats deliver have alone killed billions throughout history, more than all man's wars and revolutions combined. Rats are as deadly and fecund as germs; a single pair can potentially produce 359 million heirs in three years. Driven vandals, they must keep destroying or die, their lower incisor teeth growing a full four inches a year and liable to push up and pierce the brain if not filed down by constant gnawing. Cannibals that eat their own young, the insatiable legions of them already destroy fully 20 percent of human food supplies every year, enough food to eliminate hunger on earth; they spread scores of diseases, bite millions of people each year, cause billions of dollars in property damage.

Long regarded as the insidious Judas-symbol of evil, the furtive rat has already changed the course of human history and done more harm to man than any animal but man himself, though it might be argued that rat predation is in large part the price man pays for despoiling nature himself. An eminent biologist has pointed out the obvious ways in which rats so well resemble humans: ferocity, omnivorousness, adaptability to all climes, migration from east to west in the life journey of their species, irresponsible fecundity in all seasons, with a seeming need to make genocidal war on their own kind. Utterly destructive, both species the quintessential beasts of prey, neither of much use to any other species, taking all other living things for their purposes. No wonder there are those who fear that rats will usurp man in the future—after an Engebilike nuclear holo-

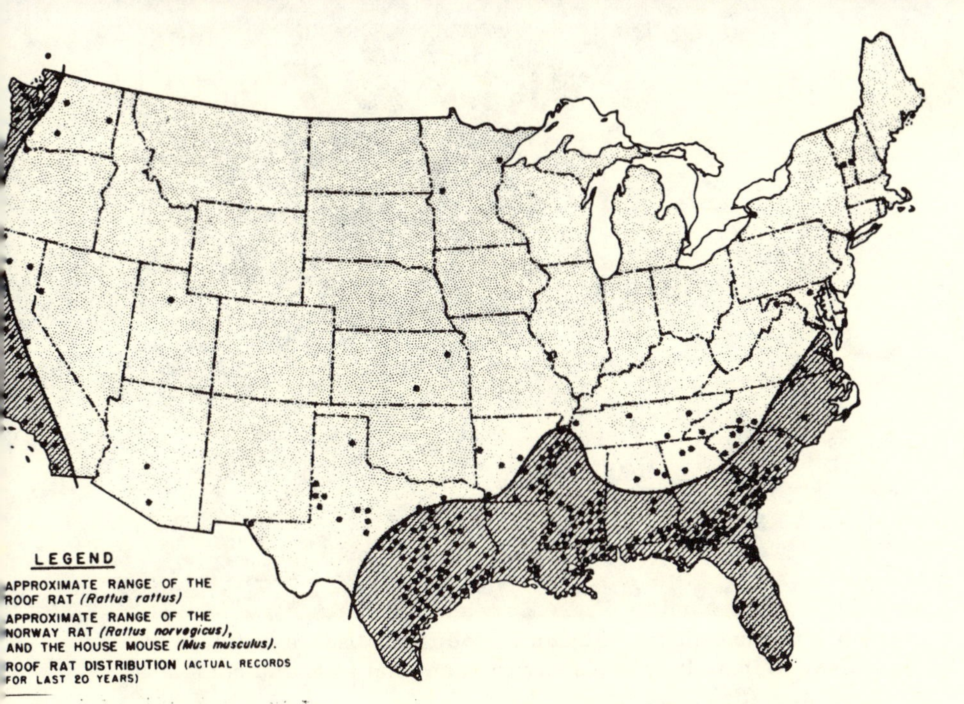


(above) Distribution of domestic rats and mice in the United States *(U.S. Public Health Service)*

JUNE 7, 1908. * *

# TS MAY DESTROY THE HUMAN RACE

**Says Dr. A. Calmette, a French Scientist—Man Must Drive Out or be Driven.**

**A MIGRATORY RODENT**

**ertion Made That a Single Pair Will Ordinarily Multiply in Two Years to Fifteen Hundred.**

ial Correspondence THE NEW YORK TIMES. ARIS, May 27.—Rats as a menace be- which humanity may disappear is a me developed with disquieting precis- by Dr. A. Calmette, a French scien- in the current number of La Revue Mois. Dr. Calmette predicts that man- d will have to engage in a general fare on rats before many more years pse if the world is to continue to be itable. He points out that different ntries have different breeds of rats ch are no great menace in themselves, ch, in fact, are often useful. The peril es from the migratory rat, otherwise wn as the sewer rat, which has been lved by civilization and which follows march of man into every clime. Rats other breeds have been known ever ce man began to keep records of the gs around him. The migratory or er rat is modern. The first mention him was made only in 1620, when he s a native of Persia and East India. did not invade Europe until the eight-

(left) *The New York Times* carried this dire prediction on June 7, 1908.

(top) The domestic rat, filthy and ridden with disease, is naturally a clean animal. It is man's environment that soils and contaminates him. *(Bell Laboratories, Inc.)*

(below) Its filth and fecundity make the Norway rat so destructive. *(U.S. Fish and Wildlife Service)*

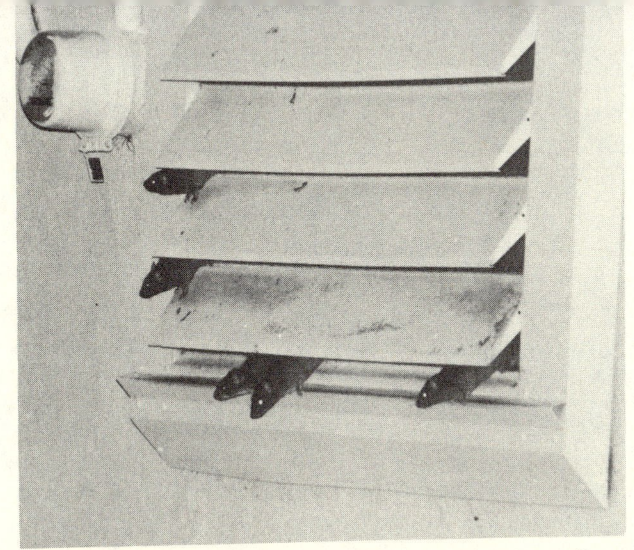

On March 6, 1978, Miami police spotted swarms of rats "peering out the louvered windows" in a bungalow. Inside, a "petite, well-kept woman" fed lettuce to the rats. "The rats don't bother me," she said, "you people do." After convincing the "Rat Woman" that the rats should be eliminated—there is no law against living with them—exterminators went to work. "I couldn't believe it," one worker said. "We killed *500* rats in one tiny house, it's incredible." *(Miami Rodent Control Project)*

(left) A Norway rat enjoying a midnight snack of cookies *(U.S. Public Health Service)*

(right) The Norway rat's need to file down its long, fast-growing incisor teeth is the reason for much of its destructiveness. *(U.S. Public Health Service)*

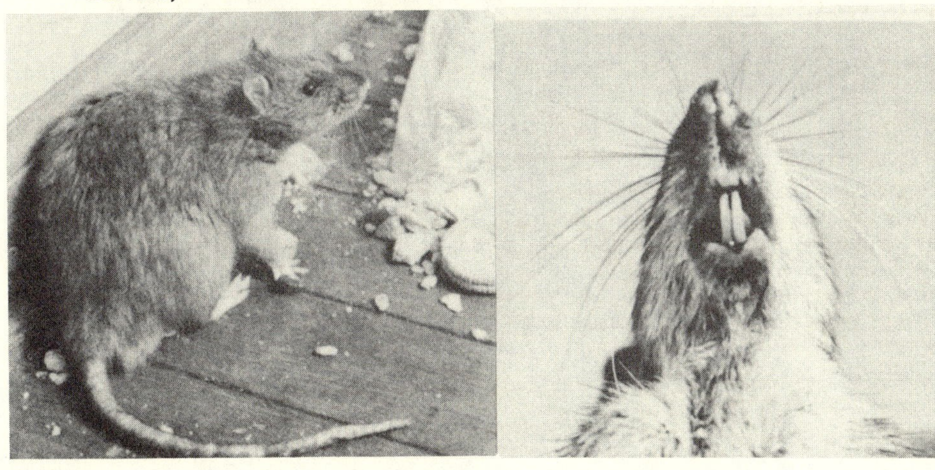

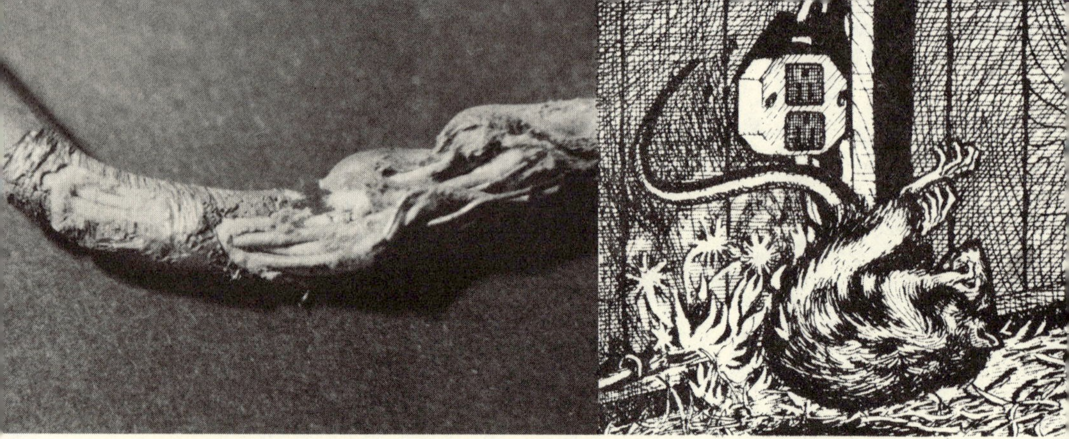

Rats cause a large percentage of fires of "undetermined origin" by gnawing insulation from electrical wires. *(Photo: Miami Rodent Control Project; drawing, U.S. Public Health Service)*

Rats bite hundreds of infants every year in the United States alone. *(U.S. Public Health Service)*

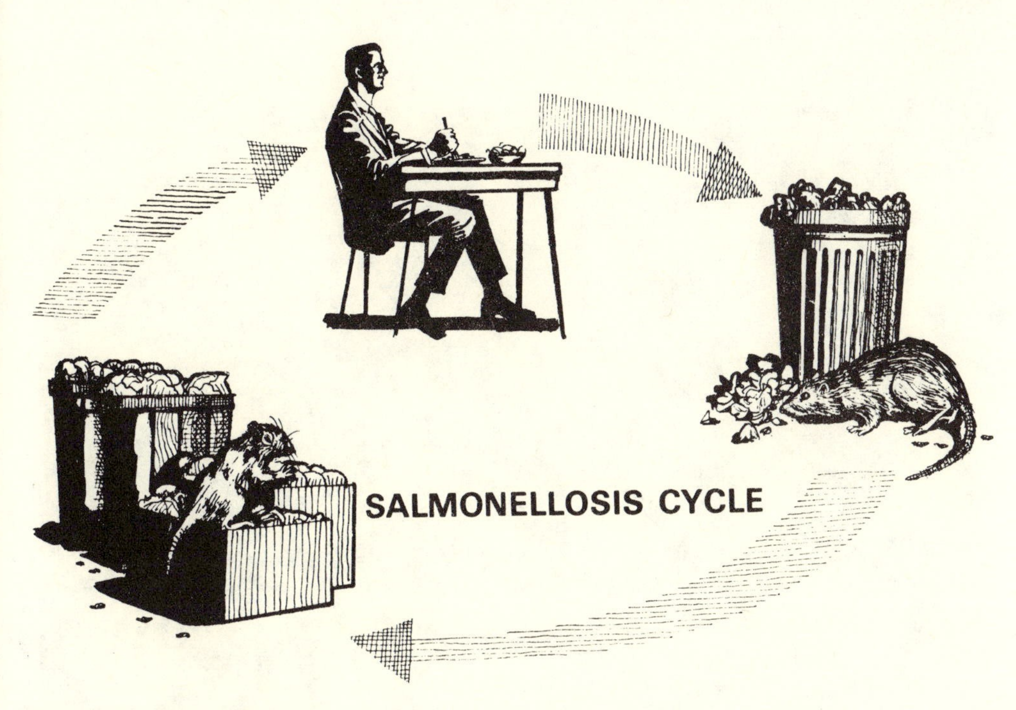


*(U.S. Public Health Service)*

A rat king composed of ten rats with their tails knotted together as shown in Henri Coupine's *Strange Animals*, published about 1850.

Naturalist Erasmus Francisci wrote about rats falling from the sky in his compendium of wonders *Der wunderreiche Überzug unserer Nider-Welt,* published in Nurenberg in 1680, as this illustration from the book shows.

caust, for example, when underground in their burrows many more rats than people would survive and ultimately conquer with dread diseases what remained of humankind in a war-ravaged world.

Such fears may have plagued a helpless woman attacked by a huge swarm of rats not long ago on the margin of a great American city. Were the voracious vandals of the night coming, were the armies of rats finally emerging everywhere in full force to vie openly for supremacy on earth? Now in delirium, now with crystal clear thought, perhaps, the woman may have believed they were as the rats inched forward. A world without people. . . . Yes the rats might be coming for all time, the Rat Age might already be upon us. . . . In any case, whatever her thoughts, the rats were edging toward her as she lay there that evening: their nerves as if raw and exposed, on the brink of hysteria, sniffing, peering here and there, bodies quivering, encrusted hair on end as they sullenly slunk along fearing all about them in the refuge of their dank burrows. The walls, the ceiling above, were alive with the sound of them, a low but distinct slithering of rats across their greasy runways, a sound that became a more pronounced scraping, scurrying and chittering, the army of vermin causing a restlessness in even the great tawny cat at the other side of the room.

Here was their eternal home, the filth of man. It seemed concentrated in this room. They boldly entered the room from a dozen different ratholes, their slick, pliant bodies squeezing through. This was always their dining hall at night, but now they sensed that the helpless woman was dying. They were ready. They waited, a lean ravenous army of giant rats up to two feet long, rats so huge and so many that the great feline trembled as it turned to confront them. But not for long. Baring its teeth, a brawny rat sprang at the cat, twisting every which way, snapping and clawing, savagely and almost blindly opening several wounds in an instant. The great cat turned tail and leapt out the window.

The woman knew they were coming, but she lay stock-still, exposed and defenseless on her bed. On this waterfront street they called her the Bird Lady, but she had long ceased to care about or even hear what the hated people called her. Even now she cared only for her birds. She glimpsed blackbirds outside perched upon electric wires as if notes or lines of music for her death song; weary of day, a white gull wheeled high and pinned a starfish in the sky to hold in place night's curtain. Weak and vulnerable, this pathetic woman had long lived to heal and feed the starving birds of the city with wastes she scrounged from garbage cans and gutters; whatever slops she did not dispense were strewn about the room, fulsome mounds of food rotting anklehigh in corners. Here in her room was the tide-wrack of life, impossible to describe or enumerate: a heap of oranges blackbearded with mold; half-eaten apples; meat and eggs splattered with ketchup and mustard; stale bread and rolls—a fermenting stew of wastes, inconcinnous debris wallowing in pools of noxious liquids. A terrible dead dove, a revolting thing with beady black eyes, lay centering it all, the bird incrusted with blood, brain, and curdled vile-smelling milk, its broken wings spread as if it had been trying to soar free from the ejecta in which it was caught.

This room, this world of hers, was the rats' world now. The pestilential odors emitted from the room, shreds of skin, bone, food, human wastes, rotted entrails, steamed up as if from an exhumed mass grave and the rats moved forward toward the uncovered once-beautiful woman, her skin now as yellow with cancer as the mustard walls, her eyes black and bright as theirs, her long raven hair a cave for eyes glaring from out of the past.

She raised her hand weakly, but they attacked, more of them slipped from their dank burrows, they covered her, a moving mound without beginning or end feasting upon her, her bony fingers barely protruding from that sleek velvet hill of them, that tumulus spotted with scabrous wormlike tails and darting eyes that flicked and blinked and swept the room like all the stars gone mad as they attended their gruesome banquet. The rat, a commensal animal, one that shares man's table, was now burrowing inside and consuming its host. . . .

Though the rats of course were not coming, not that night just a few years ago in a major American ratropolis, they had hardly finished for the evening. Throughout the world there were hundreds more such attacks. Only two miles or so farther uptown in a tenement section of the same city, other rats, huge so-called super rats immune to many modern poisons, watched in a littered alley as a helpless drunk stumbled, hit his head and collapsed in a pile of trash. The rats ate him alive before he regained consciousness. Several hours before this, a Brooklyn man was found dead in his room by police who said rats had devoured every bit of flesh on him: "His eyes, nose, ears, fingers, genitals, nothing but bones left," a detective reported. "They stripped him." Much the same happened to a man who got lost in an abandoned Pennsylvania coal mine. During the night, rats attacked him and amputated his legs and arms.

Unfortunately, such stories aren't rare occurrences. Every year, infants are gnawed to death by rats. Rats have often crippled children for life by gnawing their toes or fingers to the bone. In Detroit, an infant was bitten to death in her crib by rats that one veteran exterminator called "domesticated," that is, they were so numerous and the children so used to them that they even cuddled against the babies, nuzzling them, almost affectionately playing with them—up to a point. This is not uncommon, the exterminator says. There is a picture of the child, so terrible even in this age of Auschwitz and Hiroshima that it won't be printed here. But the story is true, as are all these stories. So is the story of the rat that swam three flights up sewer pipes and bit a man while he sat on the toilet. So is the tale of those well-to-do Chicago ladies who in the year of Our Lord 1980 lived in a house filled with countless antiques—and rats of all ages. There were literally scores of rats in the house, probably hundreds, so many that when a telephone repairman reported the Rat Ladies, as the local media later christened them, and rodent-control workers were sent to investigate, the men took one look and at first refused to enter. Cobwebs hung like drapes from the ceilings, there were stacks of newspapers 4½ feet high in some places, the house hadn't had electricity or hot water for six years. Rats had taken charge everywhere. A direct quote from a memorandum in the files of the Chicago Rodent Control Program:

> Over 4,330 pounds of garbage was eventually removed from the house. This included shredded paper, rags, garbage (rat food), and also the mattresses from two beds downstairs which were riddled with rat holes and infested with rat fleas. The sofa in the front room was in a similar condition and along with some rotting furniture was also removed from the premises. Rat feces three inches deep were observed in drawers and on top of tables and furniture. In the upper bedrooms, paper which had been shredded by rats was at least 8 or 10 inches deep intermingled with rat feces. . . .

The Rat Ladies, a woman of seventy-four and her forty-seven-year-old daughter, were actually feeding the rats three times a day with oatmeal they bought specially for them. "As long as we feed them, they won't harm us," they said. "We've got them trained." Nevertheless, the rats bit them occasionally. Rats ate at their table; burrowed inside their mattresses, couch and chairs; slept in their beds; nested in the oven; occupied their bathtub; had free run of the house; cuddled up in their laps to be petted. Not a few rats, but an entire colony. Herds of them attacked the workers, went after a television crew covering the story, chased a policeman into the street. When rodent-control workers finally got a foothold, driving the rats back into the walls, the Rat Ladies attacked. Eventually subdued, they had to be carried bodily out of the apartment away from their beloved rats.

Two years before this, the Miami Rat Woman, as local newspapers called her, claimed the rats in her house were all "just as affectionate as kittens or puppies." Eventually, over *500* rats were exterminated in her little bungalow alone. "I've been a cop for 25 years," said a police officer on the scene, "and I've seen about everything there is, but I've never seen anything as weird as this." As for the petite well-kept Rat Woman, she simply said: "It's not rats but *people* that bother me. I wish you'd all leave me alone. I pay my taxes."

But there is usually no element of comic relief in rat infestation cases. Tens of thousands of children and defenseless old people are seriously injured and killed by rats every year. Rat bite, though its significance does pale before the destruction caused by ratborne diseases, is hardly the minor problem some writers claim it to be. One expert estimates that there are some fourteen thousand cases of domestic rat bite in the United States alone every year, and this is the most conservative figure, other appraisals running as high as fifty thousand cases annually, with ten thousand more Americans bitten by rodents other than "domestic" rats each year. Rat bites worldwide must total a million or more annually, judging by official statistics like Bombay's reported twenty thousand cases of rat-bite fever every year, and 90 percent of these rat-bite victims are defenseless children. These are not off-the-wall reckonings, but conservative calculations by such respected agencies as the World Health Organization (WHO) and the U.S. Public Health Service.

One of America's first widely reported rat-bite cases concerned an unmarried immigrant girl admitted to New York's Bellevue Hospital on April 25, 1860, and placed in Bellevue's "Waiting Room," where about twenty other women slept. At that time, wharf rats controlled the world-famous hospital at night, especially in the women's wards, which they entered from the sewers and where, in one evening, forty were clubbed to death in a bathtub. Sometime during the night the woman unexpectedly gave birth to a baby girl without medical assistance. Neither did help come when she felt "rats scampering all over her, she and her newborn child bit in several places." *The New York Times* called the case "so revolting as to carry its own comment," the details "unfit for publication," but nonetheless gave the following particulars, which the editors apparently didn't find as horrible as all the rest: "At six o'clock in the morning Dr. Hadden, the House Physician, was summoned . . . and found the new-born infant lying partly under the body of the mother, dead and cold. 'The nose of the child, upper lip and a portion of the cheeks seemed to be eaten off,' said Dr. Hadden. 'The toes of the left foot and a portion of the foot were eaten off . . . The lacerated portions were covered with sand and dirt . . .' The mother . . . feeble and hardly accountable for her words . . . declared that it did not make any difference to her if the child was dead or alive. . . ."

Over half a century after the Bellevue incident, in 1926, the prominent Soviet physician Dr. Maria Baron and two nurses were sentenced to two years' imprisonment for failing to give medical attention to a five-day-old infant killed by rats in a Moscow hospital. Dr. Baron's defense was that the child had been so badly mutilated by the rats that it was doomed to die anyway and that, moreover, she was about to begin a serious operation on another patient when the infant's case was called to her attention.

Infants are still being bitten to death by rats in this age when men walk in space and take giant steps for mankind, probably four times as many as are reported, due to the social stigma attached to rat bite. In 1972, for example, a four-month-old boy was attacked and killed by rats in Alexandria, Virginia, and there have been infant killings by rats every year since then. It is much the same worldwide. From 1960–1982 confirmed deaths by rat bite have been recorded among children and adults in St. John's, Canada, where an infant died of bites to the head; Paris, where a baby died after its nose, chin and fingers were gnawed off; Turin, Italy, where an elderly man was killed; and Messina, Italy, where a young student who had taken sleeping pills was attacked and killed by a swarm of rats. Reported incidents of maimings such as noses being half bitten off would probably fill a book, and accounts of annual *unreported* deaths and injuries from rat bites might fill a set of encyclopedias. As long as such totally unacceptable horrors happen, there is no consolation in the fact that rats will *usually* run from man, or in those reports, often issued by urban health departments, that more dogs, cats, and even people bite people every year than rats do. (The 1979 New York City figures were 255 rodent bites, 871 cat bites, 973 human bites and 15,814 dog bites.)

Rat-bite stories are not products of the fevered imaginations of feature writers with a deadline, as this excerpt from an instructional guide for field workers in Chicago's Rodent Control Program illustrates: "The occasional news stories of babies or older infirm people bitten by rats will give you an insight into the vicious character of the rat. They seem to be able to sense such victims' helplessness and after prolonged contact in the same habitat are not averse to nipping some flesh from the victim. When no attempt is made to deter them, they will become more vicious and daring in their attack."

For more than a century now, the world's newspapers have featured rat-bite stories designed to shock, instruct, or both. Back in 1909, long before Harlem was a black community, a headline heralded THE BIGGEST RAT IN HARLEM. "A gray terror that bites folks and chases cats in Manhattan streets," this "monster rat" bit at least three sleeping people in an apartment house over a three-day period and finally took on a cat sent in to kill it, "the cat last seen going out of one of the windows with the rat close on its heels."

In another case at about the same time, an attendant at a hospital for the insane became an inmate after being awakened in the middle of the night by a rat which jumped onto his chest and bit his cheek. This "frightened him into hysteria" and his shouts attracted other attendants "who seized him, threw him on a bed and tightly bound him with sheets as a lunatic."

It's no wonder that the D.T.s are referred to in Ireland as "seeing the rat," or that the rat is the animal most frequently appearing in the hallucinations of alcoholics. Few are the people who can remain calm when rats touch them, or when rats are within sight or hearing, not even exterminators. "I've been in the business thirty-one years and I must've seen fifty thousand rats, but I've never got accustomed to the look of them," a veteran exterminator told *New Yorker* reporter Joseph Mitchel. "Every time I see one my heart sinks and I get the belly flutters." Another old-timer tells of encountering a rat pack in the cellar of a Brooklyn butcher shop: "I heard a noise and I turned around. My God, there were rats all over the place! On the pipes. On the floor. Poking their heads out of holes. I just plain panicked. I put one arm over my face and, yelling like a maniac, I went busting down the corridor looking for daylight. Well, those rats started jumping and squeaking and crashing into one another and screaming, and they were bouncing off my legs and I thought I'd have a heart attack. Halfway to the steps I fell down, but I never bothered trying to stand up. I went the rest of the way up the steps and into the street, breaking the world's record for crawling."

There are few if any animals so reviled and feared as the rat. George Orwell knew this when in his novel *1984* he had the totalitarian regime torture the hero, Winston, by threatening to place his face in one end of a narrow cage, while a starving rat is placed at the other end, "a common punishment in Imperial China." The rat, his inquisitor informs him, shows "astounding intelligence in knowing when a human being is helpless." Watching the enormous rodent and advised that "sometimes they attack the eyes first, sometimes they borrow through the cheeks and devour the tongue," Winston is given the choice of letting the rat eat his face or having Julia, the woman he loves, replace him in the cage. "Do it to Julia!" he finally screams. "Do it to Julia! Not me! Julia! I don't care what you do to her. Tear her face off, strip her to the bones. Not me! Julia! Not me!"

In Jerzy Kosinski's memorable novel of a child wandering in a savage world, *The Painted Bird,* the rats are only less horrifying than the people portrayed. Rats seem to slink and scamper through the pages of this *cri de coeur* as if they were characters themselves, and though a naturalist might quibble with some of the author's anthropomorphic descriptions, his account of a man falling into a bunker swarming with rats is a fitting passage to end with here, being perhaps the ultimate expression of the horror of rats attacking people:

> His face and half of his arms were lost under the surface of the sea of rats and wave after wave of rats was scrambling over his belly and legs. The man completely disappeared and the sea of rats churned even more violently. The moving rumps of the rats became stained with brownish red blood. The animals now fought for access to the body—panting, twitching their tails, their teeth gleaming under their half-open snouts, their eyes reflecting the daylight as if they were the beads of a rosary. . . . Suddenly the shifting sea of rats parted and slowly, unhurrying, with the stroke of a swimmer, a bony hand with bony spreadeagled fingers rose, followed by the man's entire arm. . . . In between the [man's] ribs, under the armpits, and in the place where the belly was, gaunt rodents fiercely struggled for the remaining scraps of dangling muscle and intestine. Mad with greed, they tore from one another scraps of clothing, skin, and formless chunks of the trunk. They dived into the center of the man's body only to jump out through another chewed hole. The corpse sank under renewed thrusts. When it next came to the surface of the bloody writhing sludge, it was a completely bare skeleton . . ."

# II

# THE RATS ARE COMING

## Rat Invasions and Depredations

The cry went up among men, women, and children as the farmers extended their hoes and scythes toward the thousands of rats stretching across the fields as far as the eye could see: "The rats are here! The rats are coming!" Thousands, tens of thousands of rats could be seen stretching clear back to the horizon, eating or destroying everything around them. In a few days they would be here in the croplands. The swarms did not often attack man himself, not intentionally, but the impoverished Filipino farmers had learned to dread these *ratadas*, or periodic population explosions of rats, just as farmers did in Burma, India, and other countries. The *ratadas* came every ten years or so, when certain bamboo species ripened and provided so much food that the rats thrived. When no more bamboo remained to eat, the rats devoured everything else in the fields, causing widespread famine and illness. The croplands would be cratered by their burrows as if bombed in a war, two rats to a square yard in places. Without government aid, the only way to stop them was to kill as many as one could and let nature run her course. One might parade the image of one's patron saint through the fields, or bang loudly on pots and pans trying to drive the rats out, or flail them with sticks, or even catch them with one's own bare hands, but time was usually the only answer—though not a very good one . . .

Invasions of rats—whether in huge *ratadas* or smaller packs—have long caused famine and spread diseases around the world; in fact, some authorities believe that if such predations could just be cut in half, human starvation would be eliminated on earth. The classic example of mass invasions by rodents is that of the destructive lemming. It may not be true that lemmings commit mass suicide by marching into the sea, a proverbial legend that even the respectable *Encyclopaedia Britannica* repeats. But the all-engulfing migrations of this ruinous little rodent, closely related to the rat and meadow mouse, are among the strangest phenomena of animal life. The Norway lemming (*Lemmus lemmus*), looking like a cross between a rat and a miniature rabbit, lives on mosses and lichens in its burrows under the snow during the winter. It is an incredibly prolific creature, females giving birth to three litters every eleven months, with up to twelve babies per litter. Young from each litter can conceive when only nineteen days old, which means that females from the first and second litters are able to reproduce within the year.

Scientists have recently theorized that the extraordinary lemming population explosions that occur irregularly in periods from five to ten years are triggered by a super hormone that appears in new spring shoots of grass. This stimulative substance (the cyclic carbonate 6-methoxybenzoxazolinone) when nibbled in small doses by the lemmings, apparently increases litter size and hastens maturation. When the lemming population explodes, the fabled migration begins, the basic drive apparently the need to find a suitable nesting area.

The pernicious lemmings begin their migration by coming down from the Scandinavian mountains, thousands of them proceeding in a straight line, a few feet separating each. They eat little on the way, but destroy everything in their path. The landscape is carpeted with them and they frequently interrupt rail and highway traffic; cars are often unable to obtain traction on the highways because the roads are so slippery with their remains. Onward the line of lemmings forges, largely male because pregnant females settle down along the way in suitable habitats. The line of lemmings traces a furrow in the earth, stripping it of vegetation while making a low whistling sound. Nothing stops it. It may go around impassable obstacles such as rocks, but will immediately form into a single straight line again. The lemmings resist any attempt to stop them, defending themselves against dogs and even men. The living line avoids human habitations, but gnaws through haystacks, levels farmlands, scoots between the legs of people at times, relentlessly marching on.

When they come to water, the lemmings, good swimmers, continue on their course, thousands of them plunging in and trying to make it to the opposite shore, often climbing over boats rather than be diverted from their straight line. Needless to say, thousands are killed along the way by man and his conveyances, by predators, and by drowning. The line doesn't steer away from wide rivers and (apparently believing they can reach the other side) even plunges into the ocean on reaching the coast, where the exhausted lemmings, *always struggling to stay alive,* drown or are swallowed up by fish and seabirds. Those females who remained behind in suitable habitats make their way back to the mountains, so this amazing mass behavior is not suicidal, as many people believe. "The false idea that lemmings have a death wish conforms to some evident need in rhetoric," columnist Philip Howard wrote not long ago in the *Times* of London. "It is all bunkum. Lemmings just don't do what they are supposed to do. The only animal that regularly commits mass suicide is *Homo sapiens.* But evidently we have a need for some vivid metaphor from nature to illustrate the human propensity to self-destruction. The poor bleeding lemming has been adapted as a cliché to fit the description."

Lemmings are only one example of the many rodents, including mice (see chapter 9) and numerous species of rats, that plague mankind in fantastic swarms. One of the great fears of firemen is the hordes of crazed rats that often attack while they are fighting a blaze in an old building. In a recent Philadelphia fire, three residents of an apartment building had to kick and fight their way down from the fifth floor when scores of squealing, snapping rats tried to escape the flames by climbing *up* the stairway toward the roof.

Invasions of snapping rats scuttering out of or into buildings or streets are not commonplace, but neither are they rare. At certain times there seem to be virtual epidemics of rat-bite cases in different areas, often for reasons unknown. HUNGRY RATS INVADE CITY—SIX PERSONS BITTEN IN BED, an early-century newspaper story begins, and goes on to tell of rats running wild on the streets, a man waking in the middle of the night to find rats under the covers nibbling his toes, an elderly couple awakening to discover rats chewing on their ear lobes, and a baby's finger chewed to the bone by rats before its mother can respond to its cries. "Officials," an accompanying editorial complains, "have not been able to assign a cause for the unusual viciousness of the rats."

In other cases the causes of rat invasions are fathomable, as when subway construction, or the razing of buildings drives rats into the streets searching for new homes, or when rodent food supplies are cut off because old marketplaces are moved. Over 230,000 rats were killed in one week when The New York City Central Park Zoo was being restored in 1934. A recent rat invasion at the Washington Monument and Jefferson Memorial in Washington, D.C., where in some areas rats already outnumber people five to one, occurred because the rat invaders had been evicted from nearby razed urban renewal sites and were attracted to the food dropped by thousands of people who visit the shrines. Another invasion in the northern New York town of Dannemora was attributed to rats' deserting a huge dump in the center of the village that Clinton State Prison had stopped using for the disposal of raw garbage. The mountain of garbage was covered with so many rats that it "appeared like it was moving" as the rats emerged from their burrows and began ranging into village homes in search of food. Downtown Newark, Chicago, Detroit, and even posh upper Park Avenue in New York City are only a few places that have experienced similar rat invasions. In some cities there are hundreds of reports annually of rats invading houses through sewer pipes leading into toilets.

An invasion of rats in 1878 caused a Brazilian famine, as had a Bermuda rat plague in 1615. Several islands have been literally overrun with rats that escaped from shipwrecks in adjacent waters and multiplied. Sea Island off Nova Scotia suffered such an infestation, the rats destroying all the ducks and gulls and threatening the lobster industry. Long before this, in 1685, rats caused the deaths of all the residents of Rona Island off Scotland when they came ashore from a wrecked ship and ate the islanders' food supplies, later feasting on their corpses.

Rats infested the former stable Napoleon inhabited during his last exile on St. Helena, biting the hand of his aide General Bertrand, attacking Napoleon's horse, and literally overrunning the place, especially during dinner, when, as the emperor's close associate Las Cases noted in his *Memorial of St. Helena,* the enormous bold rodents were attracted to the table and "it happened more than once that we had to do battle with them after dessert." Napoleon's grooms tried to raise poultry near the house but the rats "went so far as to climb into the trees to kill the fowls roosting in them." Dr. O'Meara, Napoleon's physician, wrote that "At night, startled by their sudden appearance in my room and scuttling across my bed, I would fling my boots, my bootjack, or anything else I could seize at the rats, but without the least effect, so that finally I was obliged to get out of bed to chase them away." So many poisoned rats crawled away to die under the floorboards or between the walls that they "spread an intolerable stench" throughout the house. But perhaps the worse indignity befell Napoleon himself when he reached for his familiar three-cornered hat one evening and a huge rat jumped out of it. It would not be inconceivable that Napoleon's last illness, so mysterious that books have been written debating the cause of his death, might have been one of many ratborne diseases. Anyway, his battle against the rats on St. Helena was clearly his second Waterloo.

Many are the emperors, czars, commissars, and democratic communities that have advertised for a modern Pied Piper (see chapter 7) to drive swarms of rats from town. In one such case an exterminator finally succeeded in ridding a Haverstraw, New York, schoolhouse of a plague of rats that somehow happened to descend upon it, the "slick gray rats scurrying about the building with the same confidence and almost as much dignity as the teachers, eating chalk, blackboard erasers and book bindings when they couldn't get at the children's lunches." The story of an even more unusual and alarming mass infestation of this kind at Towanda, Pennsylvania, appeared in a local paper in 1884 under the headline RATS BY THE THOUSAND:

> Several years ago a farmer living in Burlington Township, this county, received as a present from a friend in England a pair of peculiar rats. They were about one-third larger than the common mouse and their hair was a dark blue color. The farmers kept them in a large cage, where a large litter of young

ones was born. These scattered about the premises, and in a year not only the farmer's place but the whole neighborhood was overrun by the rats. . . . All attempts to exterminate them failed until a pair of pet Norway rats belonging to another farmer escaped with a large family of young, from their cage. These rats also increased rapidly and began a warfare against the little blue English rats. In a short time the latter were exterminated or driven away. About a year ago farmers in different parts of the township noticed now and then rats of an enormous size and of a breed never before seen in the county about their premises. *They were nearly as large as muskrats and of a light gray color.* They exhibited very little fear and at times boldly disputed possession of barns and outbuildings with their owners. These rats are now overrunning the neighborhood in immense numbers and have become a source of much terror to the inhabitants. They undermine cellar floors and walls and the foundations of buildings and have destroyed many cisterns and ruined milk houses for the purposes for which they are used. Many farmers have had to abandon their cellars. Granaries and barns swarm with them day and night. Farmers say that damage to the amount of thousands of dollars has been done by the pests this season.

To illustrate their boldness and ferocity several instances of recent occurrence are related. A farmer's boy entered a corn crib, in which he had discovered a number of the rats, and attacked them. They turned upon him and fought him so fiercely that he was compelled to retreat and leave them masters of the situation. He was badly bitten on the legs and hands. In their attack on the boy they sprang upward as high as his waist in their efforts to get at his face and throat. A cat, after stealthily watching four of these large rats walking about a house, finally sprang upon one of them. The other three at once attacked the cat, and fought her so desperately that she retired hastily from the conflict, bleeding from numerous wounds they had inflicted upon her with their sharp teeth. One farmer tells of a neighbor whose wife was awakened one night by screams issuing from a room where two of her small children were sleeping. She ran to the room with a light, and found that three of the immense rats had attacked the children while they were asleep, and who stood their ground when the mother came to their rescue, followed by the father. The latter killed two of the rats with a long hoe handle, and the third one escaped. The children were both bitten on the hands and in the face. The inhabitants of the neighborhood are so much alarmed by the bold and destructive incursions of these rats that they intend to hold meetings to devise some means to rid the community of their presence. Where the rats came from originally is a mystery. Some of the farmers believe that they are a cross between the Norway rat and the muskrat, which are numerous in the vicinity.

When Bertrand Russell, woman's suffrage and Liberal candidate for Parliament in 1907, opened his campaign with a public meeting in a London hall crowded mostly with women, there were those present who first thought that nonhuman rats had invaded British politics. No sooner had Russell spoken three or four words when at least forty "shockingly big" rats came scampering down the aisles terrorizing the audience. The meeting "adjourned in great disorder" and few women were present at the next one, even though several jeering men hostile to the woman's suffrage cause had been seen loosing the "rat political agents" to break up the gathering.

In their invasions, rats have been responsible through the ages for more human misery than any group of mammals, even man. Terribly damaging as it is physically and psychically, rat bite still does not cause a fraction of this death and destruction. It is impossible to accurately estimate the billions of deaths caused by their ravages, yet to say that rats have killed more people than all the wars and revolutions in history is not hyperbole but understatement.

Rats have caused more starvation than wars or crop failure. RATS MAY DESTROY THE HUMAN RACE, a newspaper headline cried seventy-five years ago. The story below disclosed that the distinguished French scientist Dr. A. C. Calmette had predicted with "disquieting precision" that unless man engaged in a "general warfare with rats . . . not many years would pass before rats would be the only animals left on the surface of the globe" because of their aggressive predatory natures. The prediction is not as fantastic as it seems at first glance. A U.S. government report concludes that "each rat damages between $1 and $10 worth of food and other materials per year by gnawing and feeding, and contaminates 5 to 10 times more." Applying these figures, which are judged conservative by some experts, to just 200 million U.S. rats, we get a figure of between 1 billion and 20 billion dollars in direct economic losses every year. Allow for only 100 million rats and the loss is between 500 million and 10 billion dollars.

Rats account for destroyed food by eating it, gnawing it, or contaminating it with their feces, urine or hair, and in their warehouse rampages they do all three at the same time. In a typical year, the U.S. Food and Drug Administration is forced to destroy over 400,000 tons of food contaminated by rat droppings, even though FDA guidelines for contaminated food make for depressing reading, allowing, for example, "an average of two rodent hairs per one hundred grams of peanut butter." Such figures do not take into account the rat-contaminated food that manufacturers and retailers destroy themselves. Or sometimes do not manage to destroy. In one outrageous case, executives of a chunky-style peanut butter company were sentenced to ten days in jail for health code violations when a 1972 Oregon health department investigation disclosed an unconscionable number of rat hairs and droppings in their product instead of the peanut chunks consumers expected. In another case, the Oklahoma Coca Cola Bottling Company was ordered by a jury to pay $125,065 in damages to a man who found a decomposed rat in a bottle of Coke he had been drinking. Five witnesses testified that in August 1979 the man obtained the bottle from a vending machine, opened and drank it and only then found the rat inside. Acid in the Coca Cola was said to have acted as a decomposing agent. The rat could have squeezed into the bottle, but how it got by the five inspectors who check each bottle of Coke on the company assembly line is another question.

Rats are so costly to Peru that that nation listed 25 million rats in its 1979 census. Deprive the rat of human food and man could erase world hunger, some experts say. A little quick math shows that if each rat destroys just five dollars worth of food a year and if a human could be fed, barely, on six dollars a day, the destruction of less than half the world's rats would save the lives of the five million people who die of starvation every year. World totals for food destruction by rats soar high into the billions of dollars, exactly how high there is no way of determining. Rats and insects, according to the World Health Organization, are responsible for 20 percent of the foodstuffs planted by man never reaching his table. We do know specifically that rats in Asia destroy over 50 million tons of rice alone every year, enough to feed one-quarter of a billion people. In southeast Asia rats consume 33 million tons of food a year, spoiling over 20 percent of man's crops before harvest. Indian rats eat or contaminate enough grain each year to fill a 3,000-mile-long freight train. In India, where there are an estimated 2½ thousand million rats, over four times as many as there are people, every six rats eat as much as one person, so that the elimination of rats would easily eliminate hunger there.

From 5 to 25 percent of all fires of undetermined origin are caused by rats, the damage certainly exceeding a billion dollars. Some fires are started by these vandals of the night filing down their evergrowing teeth on the insulation around electric wires, but many are said to result from rats stealing friction matches from kitchens and accidently striking them near piles of litter in the walls of buildings. The firebug rat might be gnawing on a match in his nest built of paper, leaves, or oily rags. The match ignites, and another fire of undetermined origin is recorded.

More than one major Manhattan blackout has been caused by rats chewing through electric insulation at a powerhouse and short-circuiting the generative system. Rats have also started floods by gnawing holes in concrete dams, dikes and water pipes. In 1945, it was reported that over a period of five years during the Nazi occupation hungry rats had chewed up the insulation of ten thousand miles of cable and consumed or wrecked about ten thousand tons of rubber in the telephone cables system, putting at least one Paris exchange completely out of business. During the Indo-Pakistan War of 1971 the Indian occupation army in the Sind Desert found that mines were being set off by packs of burrowing rats. Rats have caused damage on buses, trains, ships, and submarines. Man can't even get away from the persistent rodents by flying off into the wild blue yonder. One brown rat got aboard a Navy plane in Cuba back in 1926 and gnawed to bits everything that wasn't metal, including the parachute. Then it leaped on the pilot, nearly causing a crash, "only his steady nerves enabling him to make a safe landing."

More recently, in February 1983, American Airlines Captain Karl Burrell reported a big rat foraging aboard his plane in the first-class cabin as he prepared for takeoff to New York from Dallas. "The stewardesses prevailed on the captain to turn the plane around, and we sat on the ground for 45 minutes," said an eyewitness. "Finally, Captain Burrell polled all the first-class passengers and they decided we should go on with the rat. It was at least six inches long. We heard the captain say to the control tower, 'Let them take care of the problem in New York.'" The airline's public relations director euphemistically called the rodent a "Texas mouse," but a local exterminator said "Smells like a rat to me—we don't get six-inch mice even in Texas." The rat was killed once the plane landed at La Guardia.

Rats, rats, rats, and more rats. They have gnawed through heavy mail sacks and chewed up mail, have chewed up rare books, paintings and love letters. One man went to take his deceased wife's will from the bureau drawer where she had deposited it twenty years ago and found it chewed to pieces by rats. There have been many cases of rats chewing up paper money. Rodents often steal currency for use in making their nests. A rat nest made of seventeen dollar bills was found in Manhattan when the Civic Repertory Theater was torn down, and a Connecticut rat made a nest of nine $1 bills and one $5 bill under an old counter in a general store, the rodent gnawing a hole into the money drawer to get the money out. Think of all the money thefts attributed to innocent employees that may have been perpertrated by rats. For example, detectives investigating the robbery of a New York City A & P grocery store in 1932 never smelled a real rat. All the store doors and windows were intact, $104 was missing from the previous evening's receipts, and the manager and his two assistants indignantly asserted their innocence. The police figured it was an "inside job" but never dreamed how deep inside. Intent on clearing themselves, the three employees searched the premises after the detectives left and solved the robbery: beneath the floor were five baby rats bedded down in $90 in currency and breakfasting on the $14 remaining. A similar case was reported in Rome, where hotel employees were suspected of stealing a wealthy woman's pearl necklace from her jewel case until someone found a rat's nest under the floor boards with the pearls inside.

Rats have stolen coins, keys, lipsticks, and wedding rings. They have learned to root up turnips and carrots from the garden. A pack stole 500 carnations from a Chicago florist one night, cutting each off a few inches below the blossom before scurrying away. Rats have even been known to steal chips from a casino, and one broke the bank at Monte Carlo, or at least halted all play, when it leaped up on a table frightening all the customers away.

Rats are clever thieves, too. The old story of one rat dragging another away by the tail while it grasps an egg in its paws is examined further on, but that is far from the only trick in the rat's bag. An exterminator who hid in food places at night to observe rat behavior told this story: "There were some sides of beef hung on hooks, about three feet clear of the floor. Around 11 P.M., the rats started wriggling in. In fifteen minutes there were around two hundred in the room. They began jumping for the beeves, but they couldn't reach them. Presently they congregated under one beef and formed a sort of pyramid with their bodies. The pyramid was big enough for one rat to jump up on the beef. He gnawed it loose from the hook, it tumbled to the floor, and the two hundred rats went to work on it."

The omnivorous rat will eat anything that man eats and almost anything that man uses, from the leather gloves that we wear to the varnish on our furniture and the glue and paper in the books that we read. Rats have been seen gorging on paint, soap, shoe leather, bulbs, seeds, cloth, everything imaginable. Recently, they gnawed through and destroyed a huge batch of rat-extermination literature warehoused by a city pest control agency. They dine everywhere from luxury liners and posh restaurants, in one of which over 300 rodents were trapped some time ago, to the greasiest greasy spoons. ("People wouldn't eat out anywhere if they'd seen what I have," says one exterminator.) They are inherently wasteful into the bargain, methodically taking a bite or two from every apple or potato in a bin instead of completely eating a few. In a warehouse a pack of them will spoil hundreds of sacks of produce, far, far more than can be eaten. In one live poultry market, rats cut the throats of 325 chickens and ate only twenty-five. Another time they killed fifteen hundred baby chickens in a single foray, eating but a few score of them.

Rats hunt and eat virtually every kind of seafood, including fish, lobsters, and crabs. They have been known to cut into the bellies of pigs and calves and to dine on the oil-rich toenails of sleeping elephants, crippling them. One pack dived into an aquarium and destroyed an enormous number of valuable fish for what seemed no better reason than blood lust. The rodents have also exterminated or severely reduced many species of ground-nesting and arboreal birds, as well as small mammals, amphibians, and invertebrates. At least eighteen species of birds have been exterminated and forty made very rare as a result of rat predation, most of them insular species attacked by rats introduced by man's sailing vessels. These include the Samoan rail, the Kusaie starling, the Laysan Island rail and finch, and the flycatcher, thrush, zosterop, and starling of Lord Howe Island. Some ethologists believe the rat's ability to catch birds is an acquired skill, while others believe it is programmed genetically.

Rats surely kill as many people with the insidious diseases they carry as they starve to death. Ratborne diseases ranging from typhus to the dread bubonic plague have changed the course of human history. Ratborne typhus alone has claimed millions over the four centuries that this disease has been epidemic. In

*Rats, Lice and History*, a prestigious "biography" of typhus dealing with its impact on civilization, Hans Zinsser points out that typhus has influenced history more than any great philosopher, king, general, or statesman whose name appears in history books. Rats are the reservoir animals from which murine typhus reaches man by way of rat fleas, the oriental rat flea, *Xenopsylla cheopis*, considered the most important vector or deliverer of the disease. The causative organism, a *Rickettsia*, enters the bloodstream when a flea is crushed on the body and feces of the infected flea are scratched or rubbed into a fleabite wound or other break in the skin.

It is true that typhus can be carried by humans via lice as well as by rats, and that human carriers of typhus have usually been directly responsible for the most noted epidemics of typhus. Indeed, proof that typhus could be derived from natural infections among rats did not come until American investigators showed conclusively in 1928 that typhus cases in the southern states were infected by rat fleas from rats chronically infected with typhus *Rickettsiae*. But, as Zinsser and others have pointed out, human louse-spread typhus is the modern development, probably not commencing until the long Christian-Moslem wars of the sixteenth century in Hungary, and typhus probably originated as an ancient disease of rats and mice. In Mexico the change from rat typhus to human typhus is in progress today, there being cases of patients with typhus caused by rat fleabites creating small epidemics of the classical type. Therefore, ratborne typhus has to be blamed ultimately for famous human-louse typhus epidemics such as the one that sent Napoleon's army limping away from Russia on the most disastrous military retreat in history, or the epidemic after the Russian Revolution in 1917 when cases numbered in the millions. Ship fever, hospital fever, famine fever, military fever—the rat lurks behind every incidence of "the disease with a hundred names," including: the death of two million Aztecs in 1576; the three great epidemics that devastated Ireland; the dead bodies during the Thirty Years War that lined the roads for miles; the infected battlefields of the Crimea, where soldiers wandered delirious over the countryside oblivious to the guns and shells; the epidemics of World War I, when Serbia was effectively eliminated from the war by the disease; the epidemics of World War II, and the typhus epidemics in Nazi concentration camps, where one out of two prisoners was ravaged by the disease. No one can tell the damage it did, or the full influence it had on history.

One cannot be too careful handling not only rats, but anything that might have come in contact with rats. Ratborne diseases are spread either directly, as by bites, or by contamination of human food with rat urine or feces, or indirectly by rat fleas and mites. The hordes of rats on earth transmit at least thirty-five diseases to mankind, including another serious form of typhus called scrub typhus, bubonic plague (see chapter 3), two forms of rat-bite fever, Salmonellosis, trichinosis, leptospirosis, tuleremia, Lassa fever, lymphocytic choriomeningitis, Rickettsialpox, dysentery, relapsing fever, mud fever, swine fever, splenic fever, trench fever, Newcastle disease, Bang's disease, Chagas' disease, foot and mouth disease, rabies, listeriosis, Rocky Mountain spotted fever (which took the lives of five men from the U.S. Public Health Service who studied it), and toxoplasmosis, which can be passed from mother to fetus and cause mental deficiency in a child. Most of these diseases (many described in Appendix II ) are often fatal, and can be contacted in strange ways. Lassa fever, for instance, the most recent disease proven to be rat-borne, frequently kills Africans who eat rats. Weil's disease took the life of a New York City man who contracted it in 1977 while digging for fishing worms in earth soaked with rat urine; in another case, a student celebrating a high grade on an exam jumped into a river, contracted Weil's disease and later died of it.

Charles Darwin almost certainly had Chagas' disease, an infection of the heart muscle, which can be carried by rats. He suffered this chronic illness on returning to England from South America and by forcing him into a contemplative, less active life, it helped "to provide the almost unique combination of circumstances that allowed the idea of evolution to come to its full expression in *The Origin of the Species,*" according to one biographer. As many as seven million people in South America may suffer from Chagas' disease.

Rat-bite rabies caused many problems in Vietnam, where numerous soldiers were attacked by rats in underground tunnels that held plentiful food for the rodents. The six thousand marines at Khe Sanh were invaded by thousands of rats and all those bitten had to endure fourteen days of painful rabies shots. Rat-bite rabies has always been a serious problem on wartorn battlefields, especially during World War I, when soldiers were often forced to fight off hordes of rats in the trenches (see chapter 5) and had to hang up all food on wires at night so the rats couldn't get at it. Over one million soldiers were also afflicted with ratborne trench fever during World War I, suffering from fever, headache, rash, muscle pains, and inflammation of the eyes. Trench fever, whose germs seem to have mysteriously disappeared today, was commonly spread by rats in the trenches, the rats transmitting body lice infected with *Rickettsia* micro-organisms to the men. One field doctor gave a recipe for it that seems appropriate for all ratborne diseases: "Take equal parts of human sweat and blood and splattered flesh and bone and mix with the excrement that comes when men fear instant death. Add to the mud of the battlefields and mold into trenches. Garnish with unwashed men in filthy uniforms. Allow to bake in summer heat or freeze in winter cold and, of a certainty, lice and rats and germs will emerge to create death."

# III

# DEATH RIDES A RAT
## A Journal of the Plague Years Past and Present

The Bowers boys, aged six and seven, were laughing, playing with a dead rat carcass, as boys everywhere have done since long before Tom Sawyer twirled a dead rat on a string, much to Becky Thatcher's displeasure. But here the boys were playing by a San Francisco tenement house in a district where an epidemic of lethal bubonic plague raged. In fact, the boys were playing "funeral" with the rat, imitating the many final rites they had seen held for plague victims. Retrieving the rat from the gutter, the youngest boy handed it to his brother, who placed the rigid black rat in a small shoebox, blanketing its body with grass and flowers from a nearby lot. A service was said over the body, the boys blessed the rat, and then the mock funeral proceeded across the dirt road, the youngest pulling the box as if he were a horse drawing a hearse into the vacant lot. Soon a grave was dug with toy shovels and the rat's coffin placed in the small hole, which was ceremoniously filled with dirt and marked with a rude cross fashioned from twigs.

The boys weren't aware that they hadn't buried part of the rat. They hadn't buried the rat's fleas, which leapt from the rat's body onto theirs when the rodent grew cold, sensing the warm human flesh nearby. Several fleas had already landed on their legs and bit them, although the bites felt no different from the bites of the myriad lice they were accustomed to scratching. The boys, of course, did not know that these fleas had been infected by a rat dying of bubonic plague and had already injected thousands of *Yersinia pestis* bacilli into their bodies.

The next day, Sunday, both youngsters felt fine, even visiting the rat's grave that afternoon, but by Monday afternoon both suffered wracking headaches and complained of general weakness, soon developing chills and pain in the upper leg and groin. Their speech was slurred, tongues coated white, and their pulses rapid, while they appeared confused and apathetic, staggering when they tried to walk. The following morning their mother noticed tender egg-sized swellings or buboes on the upper thigh of one boy and in the groin of the other, where the disease had attacked the lymph nodes. The deadly *Yersinia* bacilli were already multiplying wildly in their little bodies and soon invaded nearly every organ and tissue, conquering all their bodily defenses, as happens 60 percent of the time in untreated plague cases. Their hearts fluttered rapidly but weakly, trying vainly to circulate blood through disintegrated blood vessels to the suffocated tissues. They became extremely anxious, wild with fear, then almost resigned as their skin blackened and the rigor of death set in. Two mornings later their mother found them dead in their beds. *Yersinia* the victor over their 8 million living cells, their bodies mottled with the dark spots of hemorrhage that gave the Black Death its name, the rictus of death somehow making them smile as broadly as when they were playing with the rat. Their mock rat funeral had led to their own funeral and burial in a potters field, and ultimately resulted in the funerals of their entire family of seven—mother, father, sisters, and two-month-old infant—all of whom contracted the plague from them and died within the next week.

When the Bowers family died, toward the end of America's second plague epidemic in 1906, there was still no effective treatment for the disease, although scientists had recently conquered centuries of ignorance and established the cause of the Black Death, or Black Destroyer, the "Sword of God" that had killed hundreds of millions, perhaps billions, through the ages. Strangely, despite all the horrors visited by the plague in recorded history, so far as is known no one in the West had ever firmly connected the numerous dead rats always found where plague struck with the spread of the Black Death. It was thought that rats caught the plague from men, not the other way around. Then, in 1894, six years before the first San Francisco epidemic began, Dr. Alexandre E. J. Yersin, a brilliant but unconventional Swiss-born French scientist investigating a plague epidemic in Hong Kong, discovered the rod-shaped plague organism in the swellings of victims and named it *Bacillus pestis.* At about the same time in Hong Kong the Japanese microbiologist Dr. Shibasaburo Kitasato also isolated the plague agent and he named it *Pastuerella pestis* after Louis Pasteur, which it was called until 1970, when it was officially named *Yersinia pestis* after Dr. Yersin. Yersin's discovery preceded Kitasato's by several months, but had been published in a French research journal less widely read than the English journal in which Kitasato had published his findings.

It remained for French medical officer and missionary Dr. Paul Louis Simmond to definitely establish the fact that plague was primarily a disease of rats and other rodents, specifically that rats carry infected fleas which infect humans with plague. Simmond conducted numerous experiments proving this, and in 1898 published a paper announcing his conclusions, which after some initial ridicule were tested and widely accepted within seven years. What has been called the "unholy trinity" of plague—the *Yersinia* bacillus, rat, and flea—had been established. Those few Eastern thinkers who had associated rats with the spread of the plague had been proven right, for there had been warnings in ancient Chinese, Hebrew, Arab, and Byzantine writings; the Indian *Bhagavata Parana* had advised people to flee from dying rats; and the young Chinese poet Shih Tao-nam, who was to die of plague in 1792, had written, "Few days following the death of rats, / Men pass away like falling walls." No more could the Black Death be blamed on strange planetary conjunctions, or the mysterious Plague Maiden, or on lepers or Jews who were frequently burned alive after confessions were tortured out of them.

It soon became clear that there were two kinds of plague besides the classic bubonic type. A second form called septicemic plague can kill within hours, not days; some victims go to bed healthy and never wake up, the brunt of the infection here falling on the bloodstream, which swarms with *Yersinia pestis* after an hour. The third form, pneumonic plague, springs from bubonic plague and is the most deadly of all, the most infectious of all epidemic diseases, because it can be spread without the aid of the rat flea and is mainly responsible for person-to-person spread of the disease. This form attacks the lungs and is communicable through sputum and respiratory droplets. Untreated victims of pneumonic plague rarely last three days.

Some one hundred different species of fleas can travel on the back of a plague-infected rat and leap from the deadly messenger's fur to inject *Yersinia pestis* into humans. The best-known of these servants of Beelzebub, is the Oriental rat flea, *Xenopsylla cheopis,* which is about the size of an *o*, and is wingless. It can reside by the hundreds on a rat's back, can live without food for 125 days under favorable conditions, has an excellent internal thermometer which tells it when to leave a dying rat's body, boasts a very efficient syringe for blood-sucking, and can leap so high and swiftly that for man to make a comparable jump he would have to leap over a sixty-story building at twice the speed of sound. *Cheopis* (and other similar fleas) feeds on a plague-infected rat's blood and ingests many of the plague bacilli, which kill it in time. The bacilli multiply and block the forestomach of the flea. *Cheopis* goes on a feeding frenzy to obtain food, but must regurgitate in order to swallow new blood. When it does so, whether it feeds on another rat or man, it transmits as many as 100,000 *Yersinia pestis* bacilli through its new host's skin. The ferocious *Yersinia,* whose ancestors date back three thousand million years, making them the oldest living things on earth, thirty times older than the oldest mammal, has been described as "innocuous looking," resembling "two black safety pins atop one another" under the microscope. But the bacillus is so deadly that just one of the 1/10,000th-of-an-inch bacteria is enough to kill a monkey or guinea pig. It can multiply to several hundred million in a plague victim's body.

Investigation over the years has shown that plague is harbored in over 230 animal species (many of them wild rodents) worldwide, including some 38 in the United States, which can infect man directly or transfer their deadly fleas to domestic rats that will do the job. When just one percent of a domestic rat population is infected, a major epidemic can occur. These wild animals may be dying themselves or they may be immune to plague and carry infected fleas. Ground squirrels, wood rats, marmots, deer mice, prairie dogs, guinea pigs, hamsters, harvest mice, cotton rats, pack rats, and gerbils are only a few of these rodent carriers. In addition, nonrodents such as rabbits, hares, coyotes, badgers, and bobcats are occasional carriers. Even cats and dogs can become infected by catching rats or other rodents.

Today, infected wild rodents constitute the greatest of what scientists call reservoirs of plague, which are so vast and uncontrollable that they cannot be eliminated. All experts predict that the plague will be with man forever, always ready to explode into a pandemic. The two largest permanent plague reservoirs are in the United States and the Soviet Union, which does not bode well for the world in the event of nuclear war. It also makes the prospects for sabotage frightening. Meanwhile, over 100,000 cases of plague were reported worldwide in the last decade, with probably 200,000 unreported, the rule of thumb among epidemologists being that there are a total of three cases of plague for every case recorded. The immense reservoirs of wild rodents continue to spread the plague among their own kind. In the United States alone plague has, since 1900, moved east from San Francisco at the rate of one meridian a year, leaving us with reservoirs in approximately 40 percent of the country. Some experts believe that it is only a matter of time before domestic rat populations become a vast reservoir of plague as well.

The reservoir for the most terrible of history's plagues was almost certainly today's common black rat, though the even better-known brown or Norway domestic rat was doubtless involved in more epidemics and pandemics than it usually is blamed for. Ratborne bubonic plague probably has been with us since man appeared on earth, when early hominids found it easier to catch plague-weakened rodents for food than healthy ones. But the first recorded plague appears in the Bible, where in *I Samuel* we are told that in about 1080 B.C. the Philistines were punished by the Lord after they defeated the Hebrew army at

Eben-ezer, slaying thirty thousand warriors, stealing the sacred Ark of the Covenant, and carrying it back to Ashdod near the Mediterranean Sea. The Lord "smote the men of the city, both small and great," the prophet Samuel writes, "and they had large *emerods* in their secret parts." *Emerods* probably means swellings in this sense and many scholars believe this description is of the bubonic plague, which takes its name from the Greek *bubon* for groin. After plague struck inland at Gath as well, the Philistines were directed by their priests to put the ark of the covenant and golden offerings on a cart pulled by two unguarded milk cows. If the cart was drawn back to the Hebrews at the border town of Beth-shemesh, this would be proof that the Lord of Israel had punished the Philistines. Indeed, the cows did pull the cart into Beth-shemesh, but although the Israelites exulted that the ark had been returned and made offerings to the Lord for the miracle, they too were punished with plague, for looking into the ark of the Lord: "And He smote of the people fifty thousand and three score and ten men: and the people lamented because the Lord had smitten many of the people with a great slaughter." Today, some historians think that this plague was spread by rats and men from a stricken ship in the port of Ashdod and traveled with the Philistines as they paraded the ark inland, for the Biblical account notes that there were images of "the mice [or rats] that mar the land" among the golden offerings sent by the Philistines, the ancient Hebrews using the same word for rats as for mice.

An ulcer found on the mummy of the Egyptian pharaoh Rameses V may have been a bubonic plague bubo, indicating that there was an early Egyptian plague in about 1200 B.C. The plagues of Egypt described in *Exodus*; the pestilence punishing David's sin of numbering the people, which killed 70,000 men in Israel and Judah; and the plagues that slew 85,000 Assyrian soldiers overnight, forcing them to withdraw from Judah, may or may not have been bubonic plague, but there is no doubt that the plague scurried on rats' feet throughout the world from earliest times. Herodotus tells us that plague, possibly bubonic, saved Greece when Xerxes and his invading army contracted it, lost 300,000 men, and had to retreat to Persia. Hippocrates described individual cases of bubonic plague in his work and the great Greek physician was probably familiar with epidemics of the plague, though bubonic plague may have been rare in Greece at the time due to a scarcity of domestic rats.

Ferocious plague was more responsible for the destruction of the Roman Empire than was the ferocity of the barbarians. At least eleven plagues, which may have been bubonic, were recorded by Livy in republican times, the earliest dating back to 378 B.C. In A.D. 68 bubonic plague attacked Rome, killing ten thousand Romans a day; another raging plague followed eleven years later, still another in A.D. 125, when Tacitus writes that "houses were filled with dead bodies and the streets with funerals." One can see the rats scurrying in the shadows, invisible makers of history. The great Antonine plague of A.D. 164

lasted for sixteen years, ten thousand deaths a day occurring at the height of an epidemic that exterminated half of Rome's civilian population and nearly all the Roman army, which had brought it home from Mesopotamia. The renowned physician Galen ignominiously fled bubonic plague and the emperors Lucius Veras and Marcus Aurelius died of it. Still another plague struck in A.D. 251, raging for fifteen years and killing five thousand a day at its height. Throughout the so-called "Golden Age" of Rome plague struck relentlessly, until a weakened empire fell to the barbarians. Thus travel ceased and for a while the epidemics halted.

The rat, so often associated with the devil, made a distinct contribution to early Christianity, which in the words of one historian "owes it a formidable debt." Throughout the early Christian period, as Zinsser points out, the plague that rats carried "led to mass conversions, another indirect influence by which epidemic diseases contributed to the destruction of classical civilization." Certainly many converted during the plague of Cyprian in A.D. 250. At that time the black rat can be said to have been indirectly responsible for the Christian custom of wearing black as a mourning color; the practice originated with the Roman emperor Hadrian when his wife died of plague. This pandemic lasted sixteen years, spreading from Egypt to Scotland and "all but destroyed the human race."

The first of the three greatest ratborne pandemics (epidemics that know no geographic bounds) erupted in the sixth century and is known as the Plague of Justinian because it occurred in the fifteenth year of the reign of the Roman emperor Justinian, who caught the plague but recovered. From a description of Procopius, a secretary to one of Justinian's generals, we know that the scourge was bubonic plague: "The bodies of the sick were covered with black pustules or carbuncles, the symptoms of immediate death." This terrible plague began in Egypt in A.D. 531 near today's Port Said, following an earthquake and famine. Ten years later it migrated with rats aboard Egyptian grain ships to Byzantium, today's Istanbul, where the Roman Empire had arisen again when the Christian emperor Constantine transferred the seat of government there.

The rats were to deliver a plague that left the stench of death in the air a full century before it disappeared, destroying the last vestiges of Greco-Roman civilization and killing half the population of the known world. Wrote the historian Edward Gibbon in *The Decline and Fall of the Roman Empire*: "I only find that during three months, four and at length ten thousand persons died each day at Constantinople (Byzantium), that many cities of the East were left vacant. . . ." The dead in Byzantium were so many that there was no room to bury them even when wells, caves, lakes and rivers were used as graves; the roofs of fortress warning towers had to be removed, the towers stuffed with bodies and the roofs replaced. Throughout the empire nearly half the population died, over 100 million people. The plague, according to a contemporary account, had "depopulated towns, turned the country into a desert and made the habitations of men to become the haunts of wild beasts." The rat and his black plague had helped to repulse Justinian's efforts to restore imperial unity to the Mediterranean, weaken Roman forces so drastically that the Moslems could begin their imperial expansion, and shift the center of European civilization away from the Mediterranean to more northerly lands.

There were other incidents of bubonic plague after the first pandemic. Plague turned back the Crusaders before they reached Jerusalem during the fourth Crusade, for instance, and possibly ravaged the armies of the three Crusades preceding it. But for about seven hundred years the world was mysteriously free of widespread plague, until suddenly in the Gobi Desert the rat population increased so enormously that the rats had to disperse their numbers or die. The plague rats began to migrate and finally, in the 1340s the second and most infamous of history's greatest ratborne pandemics erupted—the unparalleled Black Death.

Originating in Asia, probably in what is presently the Kirghiz Soviet Socialist Republic, then as now part of the great wild-rodent plague reservoir of central Asia, the Black Death spread with infected rats and humans to China, India, and Turkestan. Black rats were plentiful then in Europe to spread the plague, having migrated there on the ships of the first Crusaders. They had also arrived on merchant ships at least since 1291, when the Genoese navy defeated Moroccan forces and opened the Strait of Gibraltar to Christian shipping. The Black Death finally entered Europe in 1347 after a plague-stricken Turkish army besieging a Genoese trading post in the Crimean port of Kaffa (now Feodosiya) catapulted the plague-infected corpses of its soldiers into the town. The traders quickly fled for Genoa, bearing their cargoes of spices, rich cloth, jewels—and black ship rats infected with bubonic plague.

"This is the end of the world," people said frequently during the three centuries that "the Destroying Angel" waved its sword over the earth. More than 75 million people died: over 13 million in China, 25 million in the rest of the East, 7 million in Africa. In Europe at least 25 million perished, one-third of the continent's population, in the 200,000 towns and villages where the lean, long-tailed black rat spread the pestilence. Half the population of Italy died, nine out of ten in London died, only five people remained alive in the Russian town of Smolensk by 1386. The epidemic at Avignon eliminated nine-tenths of the people, the Pope consecrating the river Rhone so that its waters could be used as a burial "ground." In Rome 1.2 million people had gathered to celebrate the Holy Year in 1348, the plague spreading among them until it killed all but 10 percent. Ghost ships without crews drifted on the seas, farm animals roamed the land unattended. This was "The Great Dying," the like of which the world had not seen before and has not seen since.

The ranks of the clergy were decimated and priests deserted their flocks; able doctors like Guy de Chauliac died, as did great artists like Ambroglio and Pietro

Lorenzetti. Benvenuto Cellini barely survived an attack of the plague. At Avignon, Petrarch's Laura was taken and in Italy Boccaccio's Fiammetta, "little flame," was a plague victim. Boccaccio himself constructed his immortal *Decameron* as a collection of tales told by a merry group of Florentines who were in seclusion in the hope of avoiding the Black Death. In his introduction to the stories he gave a classic description of the pestilence, including its symptoms: ". . . it began both in men and women with certain swellings in the groin or under the armpit. They grew to the size of a small apple or an egg, more or less, and were vulgarly called tumors. In a short space of time these tumors spread . . . all over the body. Soon after this the symptoms changed and black or purple spots appeared on the arms or thighs or any other part of the body. . . . These spots were a certain sign of death. . . . No doctor's advice, no medicine could overcome or alleviate this disease."

There was indeed no cure for the pestilence, and people "falling thick as leaves from trees in autumn" often resigned themselves to death, retiring to their homes, where the doors were marked with scrawled crosses and the words "Lord have mercy on us." Yet it was amazing what people tried to effect a cure, or hold off the plague. Two lovers at Liegnitz bathed together in their urine every morning as a preventative. Eau de cologne was originally concocted as a plague protection. Menstrual blood was quaffed as a remedy. "Stinks," made of dead dogs or billy goats kept in the house, were thought to fumigate the air. Others wore human excrement for the same reason. Fearful citizens crouched over latrines all day inhaling the stench. Leeches and purges were two popular treatments. Infected men breathed their plague germs on neighbors to rid themselves of the disease. People swallowed the pus from plague buboes. Others drank molten gold and powdered emeralds. Charms, talismans, and amulets containing occult messages were believed to ward off the plague. Eighty thousand flagellants whipped themselves in penance, thousands of chorisants literally danced themselves to death on the streets. Pope Clement sat for weeks between two great roaring fires to purify the air. Probably the only good advice was the popular adage "Flee quickly, go far, come back slowly." Except that there was nowhere to flee.

Another defense, the quarantine, was invented in 1377 when the city council of what is now Dubrovnik, Yugoslavia, moved to require all travelers from an infected area to stay two stations away from the city for forty days, or *quaranta giorni*, which phrase gives us the word quarantine. But quarantine usually failed or caused more plague than otherwise would have been spread, by forcing the travelers together.

Rats often staggered out from their hiding places, reeled drunkenly on the streets and died, but they were thought to be victims of the plague, never the cause of it. However, almost everyone and everything *except* the unholy trinity of rat, flea, and *Yersinia* were blamed for the Black Death. Plague came from poisonous clouds or miasmas arising from the earth, some said. Conjunctions of the planets, volcanic eruptions, earthquakes and comets were blamed, and the Pope excommunicated one comet. Dogs and cats (but never rats) were killed for spreading the plague. Drunkards, gravediggers, strangers from other countries, beggars, cripples, gypsies, lepers, and Jews were all tortured and killed as scapegoats.

Jews found the Black Death a special tragedy, so often and viciously were they persecuted. It was whispered that the Jews of Toledo had sent agents to every Jewish community in Europe with supplies of poison made from "Christian hearts, lizards and basilisks" with instructions to drop these plague-causing concentrations into wells and springs. Though Pope Clement and Emperor Charles IV denounced such charges, few listened. Jews were tortured into confessing that they had poisoned wells or performed black magic. There was no escape. Some converted to Christianity, only to be tried for such crimes allegedly committed while still Jews and put to death horribly. Most often they were burned alive. At Mayence twelve thousand Jewish men, women, and children were thrown into a huge bonfire. The massacres were as savage and barbarous as any persecutions in history up to the time of Hitler. In Vienna the future seemed so hopeless that all the Jews locked themselves in their synagogue and killed themselves on the advice of Rabbi Jonah. Tens of thousands died and 510 Jewish communities were exterminated in Christian Europe alone, while in other places only one Jew out of five survived the Black Death pogroms.

Meanwhile the rats darted and staggered through the streets for centuries with no one suspecting them. While the Jews and other "accursed of God" were burning, the rats were stealing onto grain ships and scampering down mooring lines carrying the plague into seaports: there was no quarantine for them. Norway, all of Scandinavia, and even Greenland were struck. Throughout England children were playing funeral and singing songs like "Ring around the rosy / Pockets full of posies / Ashes, ashes / All fall down." The macabre aspect of this children's song is that it is a reference to the black plague. One of the first symptoms of the disease is a round red rash ("ring around the rosy"); friends often put flowers into a victim's pockets to make him more fragrant in his last days ("a pocket full of posies"); and finally, in the terminal stage of pneumonic plague, the victim sneezes repeatedly ("ashes, ashes" being a child's approximation of the sound) before falling down dead.

Not until the eighteenth century were the black flames of the second pandemic extinguished; part of its grand finale was the Great Plague of London in 1665, which claimed 70,000 of the city's 460,000 people. As Daniel Defoe wrote in his classic *A Journal of the Plague Year,* the plague did most damage in the filthy crowded tenements of London, where the rats were (though Defoe didn't note that), and it was popularly called The Poor's Plague, not the Black Death. In Defoe's work, however, there is a curious passage that may connect the plague

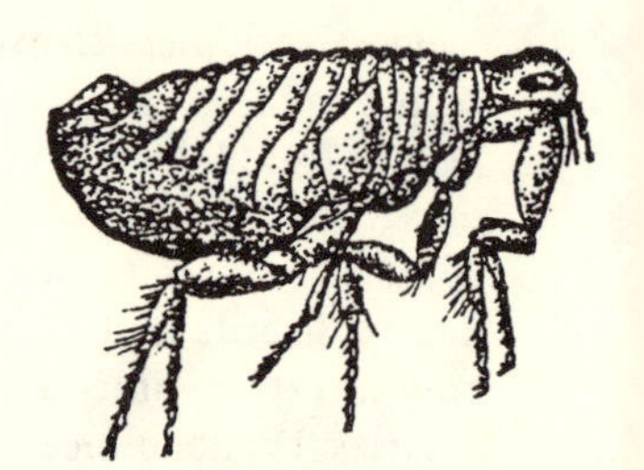

(left) The black or roof rat *(Rattus rattus)* probably brought the Black Death to Europe. *(Richard Lydekker, The New Natural History)*

(right) The rat flea *(Xenopsylla cheopis)*, which transmits plague from animal to animal, or animal to man. *(U.S. Food and Drug Administration)*

"Knock, Devil, Knock" reads the caption of this medieval woodcut depicting the arrival of the plague. *(U.S. Public Health Service)*

Accused of spreading bubonic plague, Jews were burned alive in medieval Europe, as these contemporary woodcuts illustrate.

This old print imaginatively depicts the "dance of death" performed by dying chorisants afflicted by the plague.

This etching by Daumier, entitled "The Quarantine" shows a plague house marked with a cross.

This seventeenth-century poster shows a "dog slaughterer" killing every dog he meets in a London street and a "ratter" carrying away a wheelbarrow full of dogs, which were thought to be plague carriers.

A seventeenth-century plague poster shows bodies of plague victims being buried in a churchyard. When epidemics ended, the level of many a graveyard was often a foot or more higher than it was originally, so many had been buried.

## ORGANS OF DISSECTED RATS

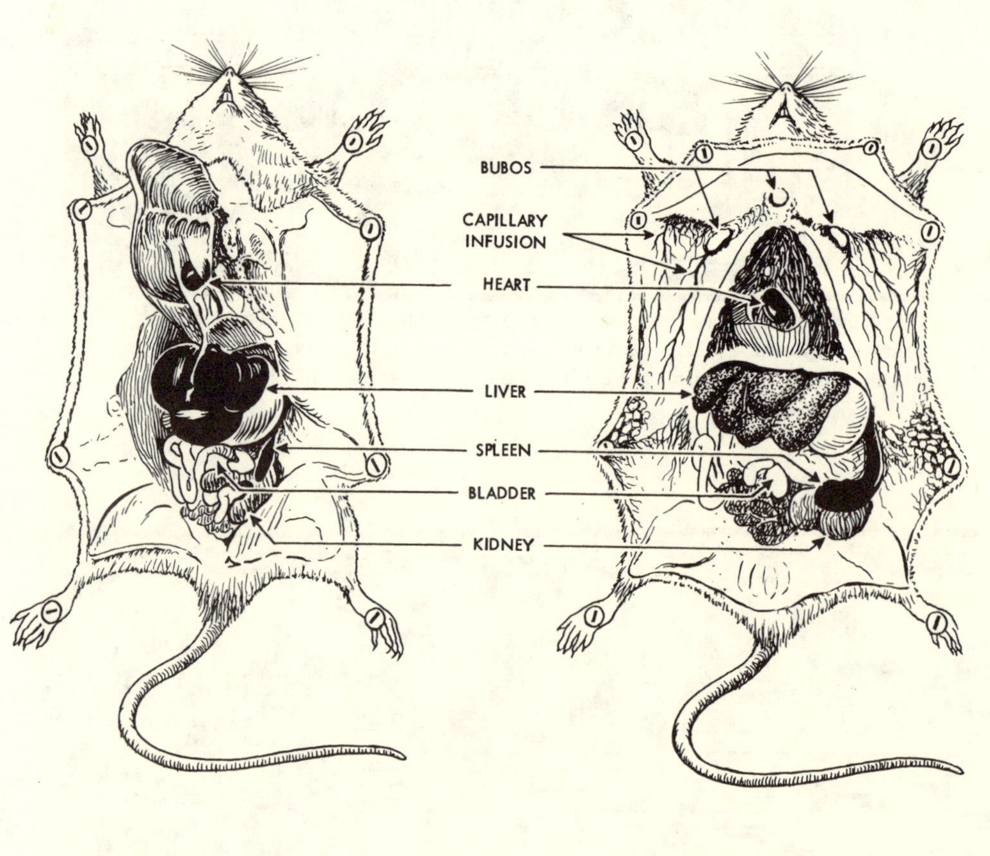


HEALTHY RAT

RAT WITH PLAGUE

*(U.S. Public Health Servic*

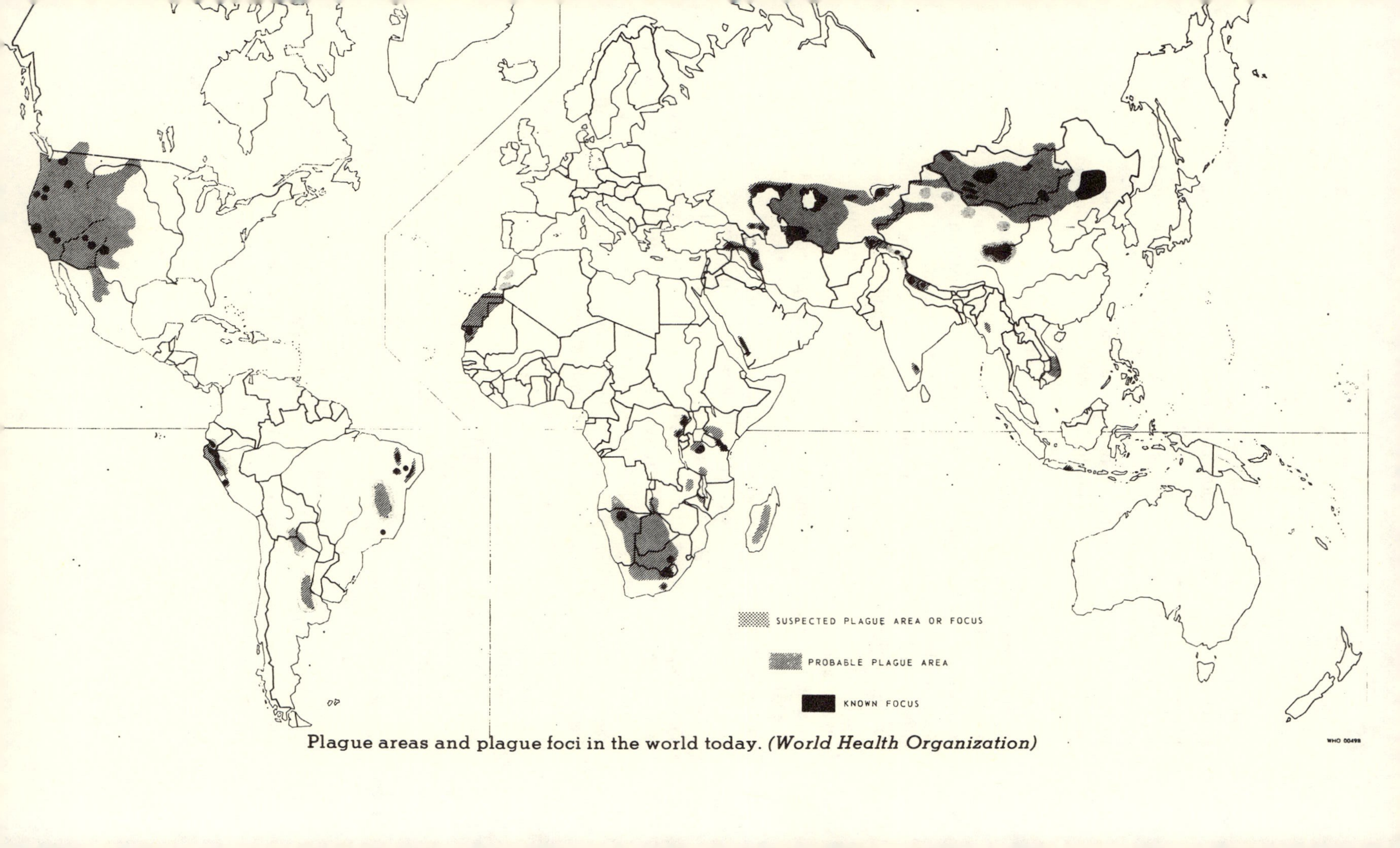


Plague areas and plague foci in the world today. *(World Health Organization)*

with rats, though the connection is rather vague and has not been mentioned by plague historians. Talking about cats and dogs exterminated in London, Defoe explains that they were thought to carry "the infectious steams of bodies infected even in their furs and hair." He then adds that "All possible endeavors were used, also, to destroy the mice and rats, especially the latter, by laying ratsbane [arsenic] and other poisons for them, and a prodigious multitude of them were also destroyed." Clearly, the implication is that rats as well as dogs and cats were thought by some to be plague carriers at the time, this an earlier reference to that fact, in Europe, than anyone has until now observed.

It is often said that the Great Fire of London in 1666 burned out the *Yersinia* infection and ended the European Black Death, but this is not true, for the fire didn't reach the London areas where plague was worst and the rats and their fleas could easily have fled the flames if it had. It may be that Norway or brown rats entered Europe driving out the black rats at the time (see chap. 3). The brown species is fully capable of spreading the plague, but it is a shyer breed that tends to live in burrows rather than close to man in his habitations. Perhaps this great plague just burnt itself out as have many bubonic plagues, no one really knows. There were several minor outbreaks, including the death of thirty men on Captain Kidd's pirate ship the *Adventure* in 1696. Then the Black Death struck in force one last time at Marseille in 1720, killing 40,000 of 90,000 people, and the next pandemic of bubonic plague did not occur for another 135 years.

One historian, G. M. Trevelyan, has gone so far as to say that the year of the Black Death, 1348, was "the year of the conception of modern man," and felt it was as significant as the Industrial Revolution. Time and intensive research have tempered or modified such judgments for most modern historians, but there remains a great deal of truth in them. The impact of ratborne Black Death on history worldwide has never been, and probably never can be, fully assessed. In its time humanity underwent a total revolution, society became disorganized, earth was delivered over to a chaos of terror, superstition, crime, pain, and megadeath in the full sense of that modern term. No society suffering a loss of one-third of its population could function effectively, military planners tell us today. So it was in the world of the Black Death. The foundations of society were gnawed from under it by rats. Scholasticism gave way to superstition, religion declined with the dearth of priests, too many of the remaining religious became monomaniacal about sin. The terrible times, including the reprehensible conduct of many priests, and what many considered the failure of mother church, helped create a psychological climate that made the Reformation possible. With the coming of the rats, books, paintings, and sculptures were left unfinished and great architectural edifices like the Duomo in Siena were never to be completed; so many skilled craftsmen died that the fine detailed architectural style called "Decorated" was lost forever. Many courageous people faced the plague and died, or emerged the greater for it, but into the world came monsters like Werner of Urslingen who wore a medal engraved with the words "Enemy of God, all charity and mercy," and cannibals like Joachim Burghard, who ate his own sister and died because she had the plague. There was a breakdown in morals unseen since the Roman emperors, whose morality was of course also influenced by plague; looting and orgies were commonplace, peasants reveled in the castles of their dead masters, drunken orgies were the way many nobles chose to live before the plague inevitably scythed them down. Soldiers refused to fight and all military activity ceased.

The most corrupt governments known to man were born in the years of the rat, and social order was almost destroyed by the deaths of so many judges and other public officials. Thousands of villages were deserted and grass grew in the streets. The economic effects of the plague were felt in every vocation from agriculture to the fisheries. Politically the plague helped foster crises like the famous Jacquerie of 1357 in France, when the peasants rose against the aristocracy, and the Peasants' Revolt of 1381 in England. The invisible rat did inspire improved sanitation systems throughout Europe, and the death of so many learned scholars forced a democratic change from teaching in Latin to teaching in the day-to-day languages of the people. But the plague's main contribution for good was its acceleration of the end of serfdom and feudalism (which the first pandemic had helped bring about). The manorial system now broke down in large part because the shortage of manpower caused by the Black Death gave those workers remaining much greater bargaining power; they could become free laborers, earning the best price for their labor at all times.

Before the rats delivered the third bubonic plague pandemic, there were a number of epidemics and smaller outbreaks around the world. Napoleon, for instance, is seen courageously touching the buboes of his plague-stricken soldiers in the famous picture by Baron Gros, "Les Pestifères de Jaffa," hanging in the Louvre. In truth, says at least one biographer, Bonaparte ordered the men poisoned so that they wouldn't spread the plague, though he surely would have done better by poisoning the local rats.

The third pandemic probably began in the 1850s when Chinese troops were sent to put down a Muslim revolt in what is now Kwangyung province, where it was said millions of sleek rats emerged from their burrows and danced until they died and their infected fleas left them. In any case, black plague among the people there slowly spread with the rats, refugees, and ragged soldiers to the seaports of southern China where in 1894 it killed 100,000 in Canton, claiming 100,000 more two years later in Hong Kong. There was no stopping the disease now. Ship rats and their fleas eventually carried black plague to every seaport of consequence in the world. It was most terrible in India, where Chinese and Indian rats rubbed noses and exchanged fleas. There millions died in the first five years, at least eleven million over a twenty-five-year period.

Ship rats carried the plague to Hawaii by 1899 and that same year finally reached the U.S. mainland aboard the Japanese ship *Nippon Maru,* which had among her cargo at least two human plague cases, stowaways, and plague-infected rats. The rats climbed or swam ashore, as two of the stowaways probably did, and about nine months later, on March 6, 1900, ironically enough in the Chinese Year of the Rat, America's first bubonic plague case was confirmed.

The Chinese laborer found dead of plague in the basement of the Globe Hotel, on what is now Grant Avenue in Chinatown, was the first of thousands of cases reported in America throughout the century. But many community leaders and state officials would not admit that plague had struck their fair city. In one of the most disgraceful affairs in the history of public health, California Governor Henry T. Gage, San Francisco business interests, and local newspapers used a score of wily strategems to block the efforts of responsible city officials and scientists who were attempting to take precautionary measures against the plague. Gage even fired members of the State Board of Health who opposed him. The plague did not exist, these purblind politicians said, because they said it did not. They ignored all the evidence. Eventually reason prevailed, but not until three years later, when a new governor took office, and after 121 more people had died of the plague, most of them needlessly.

Such conduct during a plague crisis is of course not confined to America. "One of the troubles in China," said Chinese public health officer, Dr. Wu Lien-teh, after the San Francisco affair, "is that often when the plague breaks out in a locality, the people try to conceal the fact." In one little-known isolated incident in Paris, a young woman left her hotel to pick up medicine that the house physician had prescribed for her sick mother, who had just arrived from a world cruise. When she returned from shopping, she was told that she had never been a guest at the hotel, that the room she insisted was hers long had been occupied by someone else, and that the house physician had never laid eyes on her mother or her. She returned to England without her mother and perhaps bereft of her sanity. Only years later was it revealed that, unknown to the young woman, her mother had come down with bubonic plague contracted in India, and that the hotel had conspired with police and government officials to protect France's $27 million investment in the 1900 Paris Exhibition being held at the time. Authorities, afraid of a plague panic, even went to the point of redecorating the hotel room to disguise it while the young woman shopped for the medicine. This incident, the facts slightly altered, became the plot of the movie *So Long at the Fair* (1950), starring Dirk Bogarde and Jean Simmons.

Almost as soon as the San Francisco plague struck, U.S. Surgeon-General Walter Wyman was to say, "It should not be forgotten that the epidemic is surely, though slowly, extending, and that for the first time in history it has invaded the Western Hemisphere." These were to prove prophetic words, but San Francisco learned from its mistakes and when bubonic plague struck again after the earthquake of 1906 that devastated the city (though some historians consider the years 1900–06 one epidemic), public health officials were ready. Chinatown, where all but two of the deaths had occurred, had been cleaned up and well over one million rats killed in the two years preceding the earthquake, or the plague would have been far worse among the twisted ruins. Now the U.S. Public Health Service was called in and antirat warfare resumed. A house-to-house inspection was made for rats; a ten-cent bounty was paid city exterminators for each rat caught; a new sewer system was built; free sanitary garbage cans were given to city residents; and a plague hospital was built, among other measures. The city emerged from the plague with 88 deaths in 205 cases, far better than the 102 deaths out of 105 cases in the first epidemic, and remarkable considering the conditions prevailing at the time.

Public Health care in the United States became better every year after the San Francisco epidemic, but unfortunately the attempted coverup and mismanagement of the first plague outbreak in the United States contributed in great part to the spread of plague to other states, including Washington, Texas, Florida, and Louisiana. More importnat, it allowed the plague to spread into the rodent population of California and beyond. Rats were still mainly responsible for spreading the plague. In many cases, like that of a restaurant owner in New Orleans, people actually contracted plague by picking up dead rats, but a new deadly precedent was being set. News of it first came in 1908 when it was established that an Oakland trapper came down with the plague and died after returning from a ground squirrel hunt. Investigation proved that California ground squirrels had somehow come into contact with rat fleas and that now a large number of these rodents were plague carriers as well. All attempts to eradicate the infected squirrels failed, though over a few years millions of ground squirrels were killed, and they not only remained carriers, as they are today, but helped spread the disease to the many other reservoirs of American wild animals previously mentioned. Ground squirrels were, in fact, responsible for the Los Angeles plague outbreak in 1924, when their fleas probably infected domestic rats and caused the worst pneumonic form of the plague ever seen in America. There were forty-one cases and thirty-six fatalities in what was to be the last major epidemic of plague in the United States.

In other places plague continued to rage. "There seems to be a general recrudescence throughout the world, particularly of rat plague," Assistant Surgeon-General J. D. Long, in charge of foreign quarantine, said in 1925. Sporadic cases continued to be reported in America, but they were prevented from becoming epidemic by enlightened public health practices and public education. There were several scares, though. One occurred in New York City during World War II. New York had experienced a plague scare before, in 1899, when the British ship *J. W. Taylor* steamed into the harbor with plague on board, though fortunately it never spread to shore. But almost half a century later the situation seemed more serious. On January 10, 1943, late in the evening, the French steamer *Wyoming* arrived in a convoy from Casablanca, an old home of the plague, carrying a cargo of wine, tobacco, vegetable seeds—and rats. Plague had broken out on the *Wyoming* in December, shortly before she sailed, and she was on the port's plague list, but New York public health authorities examined her crew and found no sign of illness. Then the overworked authorities, in the midst of checking a convoy, accepted a document from the captain certifying that the *Wyoming* recently had been fumigated and was rat free. That certificate later turned out to be worthless. The ship was permitted to proceed from the quarantine station outside the Narrows, to dock at Pier 34 in Brooklyn in order to discharge mail, and to unload her cargo at Pier 84 in Manhattan the next day. Luckily, longshoremen unloading the cargo reported the rats they saw in the hold, and on January 13 authorities fumigated the *Wyoming*, killing twenty rats. For forty-three years dead ship rats had been crushed up and tested almost every day for bubonic plague at the Staten Island public health station without *Yersinia pestis* ever being found. This time the usual concoction of rat spleens and livers was injected into a guinea pig and a few days later the animal died from the serum. The rats had bubonic plague. Men hurried back to the *Wyoming*, now moored at Pier 25 in Staten Island for repairs, and refumigated, killing twelve more rats. Teams of exterminators worked around the clock at all three piers where the ship had docked, laying traps for rats. It seemed almost certain that some rats must have escaped from the ship into the city to spread their germs to New York's eight million native rats. For ten weeks the exterminators killed rats along the waterfront with break-back traps, waiting for the lab to declare one of them a plague rat. Strict security had to be kept to avoid a plague scare in the midst of the war. But the laboratory never did find a plague rat. By some merciful stroke of good luck, it seemed, none of the plague rats had disembarked from the ship. By mid-April authorities concluded that the city was safe. As for the *Wyoming*, two days out, on her way back to Casablanca, she was sunk by a German submarine.

There were plague scares during World War II when the Japanese were accused of spreading bubonic plague among the Chinese by dropping fleas or flea-infected rats on them. The use of plague as a war weapon is of course nothing new, dating back to those corpses catapulted over the walls by the Tartars in the Crimea. In the early 1900s there had even been a "murder by plague" when an Indian killed his brother, the joint heir to a large estate, by hiring an accomplice to inject stolen *Yersinia pestis* into his arm while jostling him in the crowded Calcutta railway station. Plague had been reported several times after Japanese air raids on China, the first time apparently in Kinhwa in November 1940, when the Japanese allegedly dropped plague-infected fleas in packets of grain to attract rats. Shortly thereafter, 148 deaths resulted from the plague. Then Dr. Robert Pollitzer, a League of Nations health expert before he became epidemiologist of the Chinese National Health Administration, charged that "circumstantial evidence strongly suggests that a plague outbreak in Changteh was caused by enemy action." It was alleged that on November 4, 1941, a single Japanese plane raided Changteh flying "just over the roofs" for twenty minutes and dropping "scattered rice grains mixed with wisps of cotton rags" that were supposedly contaminated with bubonic plague bacilli. Six cases of plague were reported within a week and a number of rats were claimed to be infected. Only cold weather, it was said, checked the spread of the plague. Dr. Robert Lin, head of the Chinese Red Cross, backed Pollitzer, as did many authorities, but a Nationalist Chinese Commission headed by Dr. Chen Wei-kuei could not come up with much hard evidence to support the charges. Among critics of the report was an American editorial writer who said that if the Japanese wanted to plague-bomb the Chinese they would have released "a thousand infected rats from low-flying aircraft." He did not explain how the rats would survive the fall or how they would have slipped out of their parachutes if that was part of the plan. In any case, a Russian court did support such charges at war's end, finding twelve Japanese airmen guilty of germ warfare, and a Japanese documentary in 1976 charged that at least three thousand Chinese prisoners of war were killed in germ warfare experiments, including several in which the prisoners were tied up and plague-infected fleas in bomb casings were dropped near them from planes. There had been similar allegations by Japanese communists in 1946. At that time, Yoshia Shiga, editor of the *Red Flag*, claimed that Japanese medical corpsmen had inoculated American prisoners with bubonic plague virus in Harbin and Mukden, Manchuria, in experiments directed by Dr. Shiro Ishii, head of Harbin's Ishii Institute. The experiment headquarters were allegedly bombed to ruins when Soviet troops approached Harbin.

During the Korean War the Chinese, led by good old Dr. Chen Wei-Kuei, doing business for the Communist regime now, charged several times that American planes had dropped plague-infected rats or rat fleas on North Korean villages. On April 5, 1952, four villages around Kan-Hah were said to have been bombed with infected voles. No evidence for any of these charges was offered. On the other side, intelligence officers of the U.S. 187th Airborne Regimental Team reportedly found more than five thousand rats and mice inoculated with bubonic plague in a secret bacteriological laboratory operated since 1947 under the supervision of a mysterious Russian woman scientist. The laboratory had been deserted just before Pyongyang was captured after MacArthur's famous landing at Inchon. A civilian doctor who worked on the project in the North Korean capital allegedly led the Americans to the secret laboratory, where the animals were kept in cages in groups of from two to twenty and their furs were sprayed with a chemical that encouraged the multiplication of fleas. All but 380 of the rodents had died of starvation and none was believed to have been turned loose or to have escaped from the sealed cages. The mysterious Russian woman scientist, said the informer, had fled with North Korean and Russian officials, who abandoned the capital before it was captured.

In pre-CIA days, not one American in a million would have believed his country guilty of germ warfare. But many today give credence to charges that a Mafia yacht landed two Cuban-exile CIA agents in Cuba on the night of November 2, 1963, each of them carrying two special containers filled with *Yersinia pestis* and directions to find rats. The plot, dubbed Operation Visitation, apparently failed, the agents never returning the following night as planned. It has also been charged that there were experiments concerning the loosing of plague rats among enemy strongholds in Vietnam and Cambodia during the Vietnam War. Although President Nixon ordered the destruction of all biological weapons by 1973, there are those who still wonder about the CIA. As for the Russians, they have not even pretended to destroy their plague weapons, and the United States has absolutely no plans for defending civilian populations against such attacks, even the most modestly successful of which, according to WHO, the World Health Organization, could easily kill ten thousand people.

Although WHO declared that the third plague pandemic (which had lasted more than 100 years) was over in 1959, after some 13 million had died, there were still 100,000 cases reported worldwide in the decade ending in 1974. Vietnam, its plague-bearing wild rodents driven from their normal homes by conventional and chemical warfare into closer contact with domestic Polynesian rats and man, contributed well over 80 percent of these reported cases and actually may have had a quarter of a million cases, including thousands of deaths. The conditions were perfect for plague in a defoliated land cratered like the moon over a total area larger than Rhode Island. Because they were given plague inoculations American soldiers were not frequently stricken with the disease, but there were several cases among American troops and civilians on whom the inoculations did not work. In one case a twenty-one-year-old soldier contracted plague after participating in the demolition of rat-infested barracks near Saigon that had not been disinfected. As hundreds of rats scurried for new hiding places, the soldiers vied to see who could stomp to death the most rats, the ruddy brown fleas harbored by the rats jumping to the men. The young soldier contracted bubonic plague, but did not become ill until after he was sent home on furlough to Texas—the first bubonic plague case to be imported into the United States in forty-two years. There he soon entered a veterans' hospital, where his condition was misdiagnosed and he was treated for a hernia for sixteen days before someone discovered he had plague. Luckily, the disease didn't spread.

Close calls have been common in the plague-detection business, as just this brief account with all its "fortunatelys" and "luckilys" shows. As recently as September 1981 the U.S. Center for Disease Control published this record of a close call in its journal *Plague Surveillance*:

"Case 12 had travelled to Columbus, Nebraska, on 8/15 and on 8/19 was admitted to a Columbus hospital with fever and right inguinal adenopathy (swelling of the groin). She was treated with ampicillin and left the hospital against medical advice on 8/22. After her discharge it was learned that she might have plague and she was tracked down on Highway 81 by the Nebraska Highway Patrol and returned to the hospital in Columbus. Plague infection was confirmed on 8/28. . . ."

Another possible plague spreader was tracked down in Cambridge, Massachusetts, and still another in Seattle, among many places over the years. Like all plague victims these patients are isolated forty-eight hours to determine if the case is of the extremely contagious pneumonic type, and treated with streptomycin or antibiotics such as tetracycline and chloramphenicol. Any contacts of pneumonic victims are also quarantined and given antibiotics. These drugs have saved many lives and are usually effective, but just as certain "super rats" have acquired a resistance to poisons, some strains of *Yersinia pestis* have acquired resistance to streptomycin and other miracle drugs. Vaccine against plague, first made by Yersin in 1897, is also highly successful, but does produce a large number of adverse reactions. The protection is temporary, so that the injections, with boosters, are recommended only for those at high risk of the disease. Inexpensive routine inoculations can be used to build up resistance of individuals in a known plague area before plague strikes, as was done with GIs in Vietnam, who suffered only eight cases of plague overall compared with at least five thousand cases a year among the Vietnamese population.

Unfortunately, scientists and public health officers in the United States have been hampered over the last ten years by extensive budget cuts at the federal, state, and local levels and the development of new antiplague weapons seems unlikely at best in such a financial climate. In July 1981, for example, one of only four rodent-control labs in the United States had to cease operation because of budget cuts. The Soviet Union now has twenty-five-thousand people involved in plague research work, while the United States has about twenty-five. Plague surveillance work has been particularly hard hit, with the U.S. Public Health Center for Disease Control operating on a lean budget which has been cut by over one-third since 1972. Its *yearly* budget in the fight against an enemy that has killed more people than all the wars in history is less than one-half of what the Pentagon spends for defense every *day* of the year. Experts say that it is impossible to carry out the plague-surveillance program recommended by the World Health Organization with the limited funds and few people available. Others flatly predict that unless things change America will see far more bubonic plague than the very roughly estimated ten thousand cases and one thousand deaths experienced over the last eighty-two years.

With thousands of cases still reported every year and epidemics arising occasionally in India, Burma, Africa, and Madagascar, ratborne bubonic plague is far from dead or even dormant in the world and a fourth pandemic easily could be set off by war or famine. Neither is there much to celebrate in the United States, where there has not been a known classic case of urban ratborne plague since 1924 (though rats with infected plague fleas have been found in San Francisco and Tacoma as recently as 1971). Here the vast reservoirs of plague created during the third pandemic are firmly entrenched among wild rodents and could bring on an apocalypse at any moment, especially in the sparsely populated Rocky Mountain West, from which wild rodents may shortly be evicted en masse because the area contains an estimated 50 percent of U.S. energy reserves. It was in the Southwest that some 80 percent of U.S. plague cases were reported over the last decade, the victims primarily residents, especially Indians on reservations, but including tourists strollers, hikers, campers, Boy Scouts, hunters, ranch hands, and scientists on field trips. In 1981, a twenty-five-year-old trapper died of plague in El Paso, Texas. Bubonic plague-infected dead rats were found in August 1981 near President Reagan's 688-acre Santa Barbara ranch, where the President's favorite activity is clearing brush. Five months before that, New York City suffered its third plague scare of the century when a man, who had been bitten by a flea on a dead squirrel he had "playfully" thrown at his wife while hiking in New Mexico, was suspected of having bubonic plague. The suspicion proved unwarranted. It should always be remembered that wild rodents infected with plague are found in 40 percent of the area of the United States, in every state from the Pacific Ocean east to the one hundredth meridian running through the center of Kansas and Oklahoma, and that no state or city is free of the domestic rats that might mingle with these infected wild rodents when they arrive. Some experts say it will take no more than another eighty years for the plague reservoirs to cover the rest of the country. In fact, no one really knows how far the rodent plague reservoir does extend today, as no extensive studies of wild rodents in the East have been done.

The 1970s saw more U.S. plague cases reported (105) than any decade since 1900–10, and in 1980 there were eighteen cases, five of them fatal, a start that promises this decade will be worse. But these figures pale before those that would be tabulated if infected wild rodents driven from their homes, as they periodically are by brush fires in California, were to come in contact with rats in a major metropolitan area. It is no wonder that the World Health Organization has warned that "At all costs the rats that infest man's cities and the wild rodents must be kept apart, or we face the growing menace of explosive outbreaks of human plague."

As recently as 1972, the U.S. Center for Disease Control warned that "plague must be viewed not as a historical phenomenon, but as an ever-present threat, not only in the United States, but throughout the world." Dr. K. F. Myer, a distinguished authority on epidemics, has written that "the possibility of plague cannot be excluded" today, and the eminent French scientist Dr. M. Baltazard has warned that the relative "silence of plague must not blind us to the fact that its present positions are stronger than they have ever been; entrenched within reach of all the strongholds of modern civilization it may well be a disease of the future."

Faced with the sure knowledge of famines throughout the world that weaken men to disease (5 million already dying every year of starvation and 500 million undernourished); military adventurism and economic exploitation that despoil the environment; senseless budget cuts everywhere for public health service; the immense reservoirs of wild rodents and domestic rats in every land, and a global community where no two points are more than a few hours distant, one recalls the words of Dr. Rieux, the doctor in Albert Camus' novel *The Plague*, who muses after plague has left the town (Camus' world) "that the plague bacillus never dies or disappears for good . . . and that perhaps the day would come when . . . it would rouse up its rats again and send them forth to die in a happy city."

Perhaps the rats will be roused and millions will die again, the rats gorging on their kill in city streets. We can only hope not, we cannot be certain, despite foolish assurances that "it can't happen here." Man is much in the position of Robert Burns when he wrote his poem to a rodent:

> Still thou art blest, compared wi' me!
> The present only toucheth thee:
> But och! I backwards cast my e'e,
>     On prospects drear!
> An' forward, tho' I canna see,
>     I guess an' fear!

# IV

# THE LIFE AND TIMES OF BROTHER RAT

## A Not So Natural History

The two rodents posed for the camera, offspring of the same parents, were identical except that the "rat-mouse" was roughly twice the normal-sized mouse's height and weight. This had nothing whatsoever to do with better nutrition or characteristics inherited from the rodent's true parents. The giant rat-mouse had been created late in 1982 by scientists from four American universities, who had successfully taken a growth gene from a rat and planted it in the fertilized cell of a mouse to produce the first animal made by transplanting genes from one species to another. Such genetic engineering has enormous, perhaps monstrous, implications for the future, one of them being the possibility of a huge, cunning breed of rat-mice, ferocious like rats, but even craftier, since they would carry within them the mouse's deeper distrust and superior hiding ability developed over the ages. Carried a logical step further, it seems that genetic engineering has finally made possible those breeds of rats as big as wolves or tigers that H. G. Wells and other science-fiction writers have predicted would plague or take over the world.

For at least a century now, science fiction writers and scientists have been predicting that the cunning, prolific rat might wrest control of earth from man and destroy the human race. In any case, it is no exaggeration to say, as one prominent naturalist did, that rats, of the *Muridae* super family, are "the most adaptable mammal group on this earth," and fully capable of rising to the

# FIELD IDENTIFICATION OF DOMESTIC RODENTS

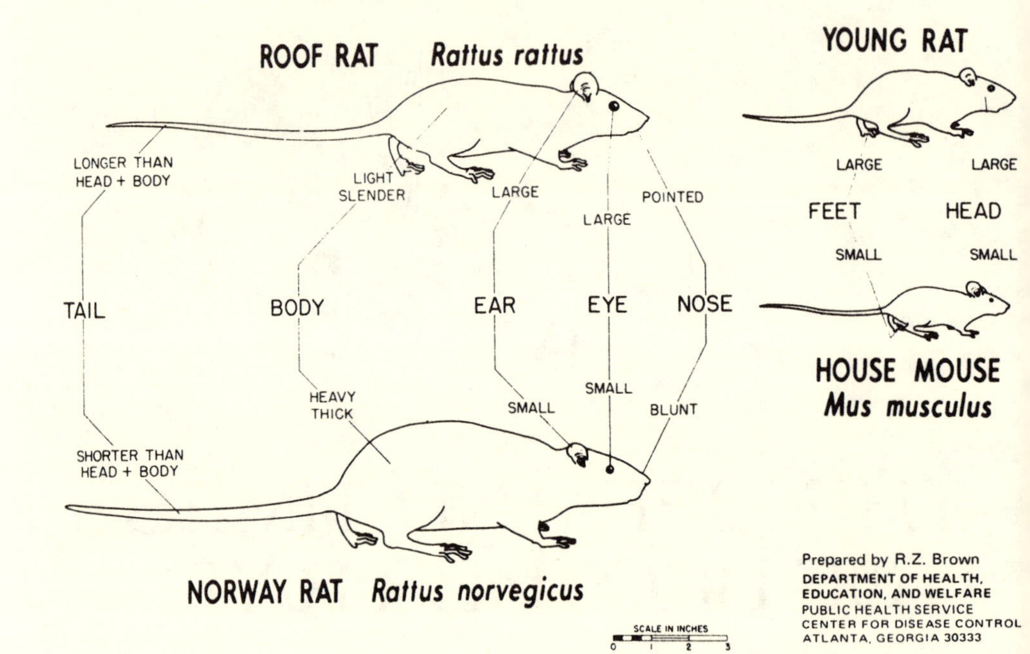


*(U.S. Public Health Servic*

| Species | Norway Rat *(Rattus norvegicus)* | Roof Rat *(Rattus rattus)* | House Mouse *(Mus musculus)* |
|---|---|---|---|
| Weight | 10 - 17 oz. (280 - 480 gm.) | 4 - 12 oz. (110 - 340 gm.) | 1/2 - 3/4 oz. (14 - 21 gm.) |
| Total length (Nose to tip of tail) | 12-3/4 - 18 in. (325 - 460 mm.) | 13-3/4 - 17-3/4 in. 350 - 450 mm.) | 6 - 7-1/2 in. (150 - 190 mm.) |
| Head and Body | Nose blunt; heavy thick body 7 - 10 in. (180 - 255 mm.) | Nose pointed; slender body 6-1/2 - 8 in. (165 - 205 mm.) | Nose pointed; small body 2-1/2 - 3-1/2 in. (65 - 90 mm.) |
| Tail | Shorter than head plus body, carried with much less movement, comparatively, than roof rat. Lighter-colored on underside at all ages. 6 - 8-1/2 in. (150 - 215 mm.) | Longer than head plus body generally moving whip-like, uniform coloring top and bottom at all ages and for all subspecies. 7-1/2 - 10 in. (190 - 255 mm.) | Equal to or a little longer than body plus head. 3 - 4 in. (75 - 100 mm.) |
| Ears | Small, close set, with fine hairs, appear half buried in fur. Rarely over 3/4 in. (20 mm.) | Large, prominent, hairless, stand well out from fur. Generally over 3/4 in. (20 mm.) | Prominent, large for size of animal, with some hairs. 1/2 in. (15 mm.) or less |
| Hind Foot | Usually over 1-1/2 in. (40 mm.) from heel to tip of longest toe. | Generally less than 1-1/2 in. (40 mm.) from heel to tip of longest toe. | Feet are shorter, darker, and broader than most wild mice. Generally less than 3/4 in. (20 mm from heel to tip of longest toe. |

*(U.S. Public Health Servic*

occasion if the opportunity presents itself. Not only are rats, which probably emerged late in Miocene times, among the most fertile and numerous of animals; they also take an incredible number of forms, far more than any other family of mammals. There are some 100 genera alone of what are conservatively labeled Old World Rats, these containing some 370 species and 1,500 subspecies.

Both the black and brown rats, which have been Western civilization's worst enemies (aside from itself) throughout recorded history, are Old World rats whose origins are Asian. They are of the true rats, genus *Rattus,* and although the exact number of their species is unknown, some 570 forms have been described. Black, or roof, rats (*Rattus rattus rattus*) are smaller and slenderer than their brown, or Norway, rat cousins (*Rattus norvegicus*), weighing from 100 to 340 grams (4 to 12 ounces), as opposed to 280 to 480 grams (10 to 17 ounces) for the brown species. They have tails longer than their bodies, while the burly brown Norway rat's tail is shorter than its body. The record sized Norway rat weighed close to 3½ pounds and measured about a foot and ten inches. The sleeker black species has larger ears and eyes, a tail with 250 transverse rings (the brown rats has 200), softer fur and a more pointed nose. Except for the lack of a bushy tail it resembles a squirrel much more than its shaggy brown cousin and, with its long tail balancing it, can climb and leap with nearly as much grace and agility as a squirrel. The accompanying diagram and charts point out specific differences between the closely related species better than could any explanation. It is interesting to note that Linnaeus dubbed the Norway rat *Mus decumanus,* or the huge mouse, associating it with the house mouse, and that it was also once called *Epimys,* or a "superior mouse."

The black rat came first to Europe and as far as we know is the rat that ferried the Black Death bacillus there, although both species can spread bubonic plague. Bones of both black and brown rats have been found in Stone Age dwellings, but there is little if any mention of either in recorded history until late in the twelfth century and their rare presence as fossils in the excavations of ancient cities suggests, that they may not have been common in the West until shortly after the Crusades, when as we have seen, the black rat came back to Europe with the Christian warriors and caused the most terrible catastrophe the world has known.

Black rats must have been numerous long before they invaded Europe, if the prevailing theory about their origins is correct. It is thought that they originated in southern Asia, specifically in South China, Burma, Malaya, and India, and that they were originally tree rats, which they are still called today because of their aboreal habits: they like to live high and dry in the attics and upper reaches of wooden houses, docks, and warehouses, and in some parts of the world still live in trees. The black rat probably plagued Asian house dwellers from the time it found it could fare better in human villages than in the wild, but little is recorded about it (at least little discovered by Occidental researchers) until hordes of black rats followed the caravan routes across India into the eastern Mediterranean region and entered Europe at the time of the Crusades. Actually, though it first became widely noticed at this time, the black rat was present in smaller numbers in Europe as early as the Ice Age and had lived in the Mediterranean area for thousands of years. Magdalenian deposits in the Frankenland Cave indicate that it was even present in Bavaria long before the Middle Ages. Probably it was responsible for spreading the Justinian Plague in 531.

The black rat reached England in the Crusaders' ships, not across the Channel from Europe, and from there took a full four centuries to reach Scotland. It only traveled the relatively short distance to Scotland after it was introduced by Spanish explorers to Central and South America in 1540 and after it came ashore with English settlers at Jamestown, where it almost succeeded in ending the English experiment in America by destroying the colonists' winter grain supply. The aggressive creature reached Australia with the first convicts shipped there and sailed with Captain Cook to the South Seas, where great armies of its kind eventually ousted the Polynesian rat, which is now found only in Fiordland National Park in New Zealand, and on a few outlying islands. Sailing to South Africa, it spread into the interior of the country, where it made its home in native *kraals*. So many places did it reach by stowing away aboard wooden sailing vessels that it became known as the ship rat, that very same rat that will desert a sinking ship. More than 90 percent of rats living on ships are still of the so-called black species.

To complicate matters, black rats are not always black, they show great color variation. Today's black rat is held by English rodent expert M. A. C. Hinton to be a cross between the gray-bellied and brownish-top Alexandrine rat (*Rattus rattus alexandrinus*) and the white-to-brown-bellied and brownish-top Sicilian fruit or corn rat (*Rattus rattus frugivorous*). Hinton's theory holds the black rat to be a melanistic mutant, or dark-color form, that was hardier in the colder climate of Britain and northern Europe where it became dominant, which may be why the black rat is not recorded as being black until about 1600, long after it had carried the plague from its Asian home to the West. Further complicating the matter is the fact that black races of the brown or Norway rat have been reported in the British Isles since the early nineteenth century. And perhaps one should not even mention that a true black rat can range in color from black to dark slate-gray! Often a single litter of black rats will contain pups of black, gray, and brown color. In 1929, a *blue* rat, believed to be a true mutation of the black, appeared in the University of Illinois breeding stock.

Under the aliases of ship rat, Alexandrine rat, Sicilian rat, fruit rat, corn rat, tree rat, and black rat—names that are popularly used interchangeably—the black rat is today found throughout most of the world, sometimes coexisting with the brown or Norway rat, but always occupying a different ecological niche, often living in the same house as its brown cousins, but taking the upstairs parts of the house as its own while the Norway rat lives in the basement and underground burrows. In its ancestral Asian home the black rat is still numerous in field and forest and retains its original coloration of white belly and rufous upper body. In New Zealand, which it reached with the first settlers and where it "moved about the country in vast armies," according to one naturalist, it has colonized the native forests and plantations. The gray-bellied Sicilian black rat that arrived in Jamestown eventually populated America's entire eastern seaboard and then moved west. But nowhere except in tropical regions does the black rat now outnumber the brown race, although some authorities say its vast populations in these areas make it the world's most numerous rat.

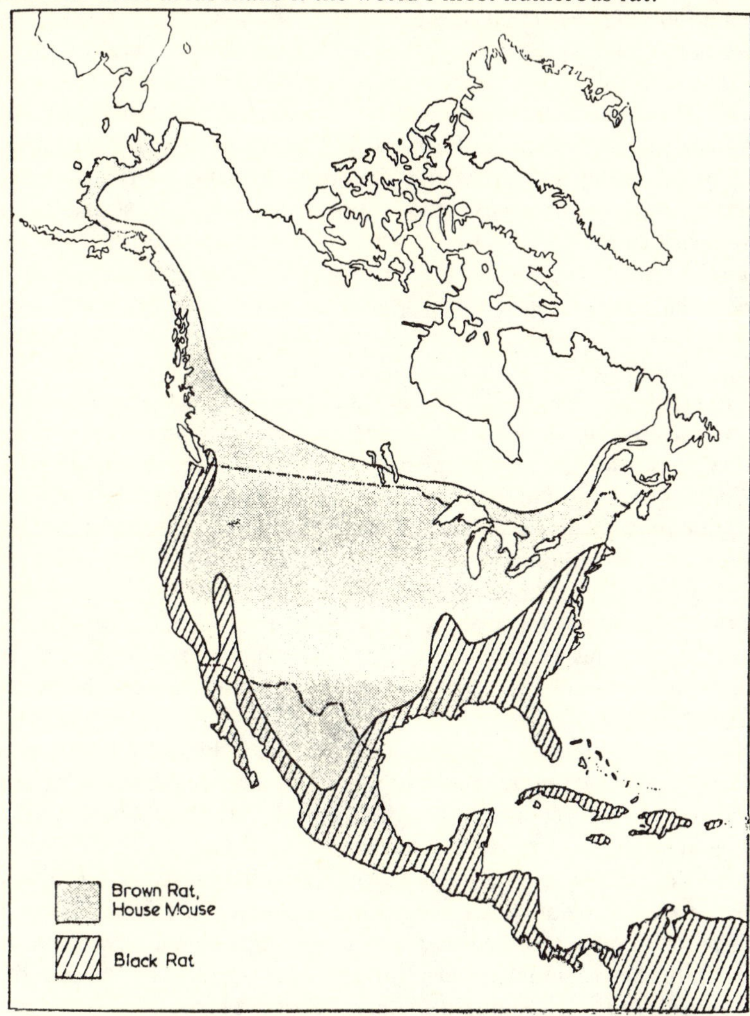


Distribution of brown (Norway) rat, house mouse, and black rat in North and Central America *(U.S. Public Health Service)*

Black rats live wild in areas around the Mediterranean; they are generally less dependent on man than the brown rats and are more frequently found many miles from human settlements. However, they do seem to be making a comeback today in England and other places because man can more effectively ratproof buildings against burrowing Norway rats than against these agile creatures. Yet they are far fewer than their rivals in Asia, Europe, or the United States, where today they are almost entirely confined to the South, in wooded areas along the Pacific Coast and into western Canada, and in parts of Hawaii. They have been trapped in the Everglades National Park seven miles or so from any occupied structures. The accompanying map shows their North American geographical distribution along with that of the brown or Norway rat.

German naturalist Peter Simon Pallas contended that in 1727, after a terrible earthquake, great masses of brown rats were seen scurrying from Asia through Russia. They crossed the Volga River, thousands of them drowning, and eventually conquered the smaller black rat in Europe. But the brown rat probably had thrived in the European wild unnoticed long before this and may have helped its black relative deliver the Justinian Plague. The first reliable observations of brown rats (except for their depiction in first century Roman statues) were made in 1730 in England, 1735 in France, 1755 in America, and 1800 in Spain. There is little doubt that there was a war between black and brown species. And what a night war it must have been in narrow, twisting corridors beneath the earth! But while it is probably true that the giant brown rats drove out the black rats in a war of "biotic antagonism" between the two species that has been called the greatest interspecific animal conflict of all time, there is another important reason for the black rat's decline and fall besides the night-long battles that raged in filthy places unseen by human eyes. For black rats also faced an environmental change they could not cope with: the gradual disappearance of wooden buildings, in which they much prefer to live because of their climbing habits. In fact, it has been demonstrated by European investigations that the black rat has managed to persevere in regions where wooden buildings have remained the most common type, and that areas such as the docks of London, Bristol, and Liverpool will provide it with city strongholds for many years to come.

Besides the physical differences noted between black and brown rats, the black rat's habits are more related to particular times than the brown rat's and it is a largely nocturnal creature. Only in tropical and subtropical areas, where no other animal destroys more food and supplies, is it a greater pest than the brown rat. For the most part it shares the same feeding habits, sexual habits and living habits; even its predators are the same. Fortunately, however, black rats and brown rats cannot interbreed, as closely related as they are. Japanese researchers have established that they could not interbreed even if they were able to tolerate each other (for they wouldn't declare a truce even for sex). The experimenters fertilized forty-eight brown rats with black rat spermatozoa. As there were no abortions and no pups were born, it is thought that the hybrid embryos died during the first half of pregnancy and were reabsorbed. The same result was obtained by other investigators and so there seems no danger that man will be plagued by a brown-black super-rat with the best or worst qualities of both.

The brown rat was christened the Norway rat by the English scientist Berkenhout in his *Outlines of the Natural History of Great Britain* (1769) because he believed it had arrived on British shores aboard Norwegian vessels. But Norway had been maligned. The truth is that Norway had no brown rats until a full forty years after they reached Britain. Britain's first brown rats probably came from Denmark in 1716 and, if this is the case, should properly be called Danish rats. One legend even has them coming to England with William of Orange in 1688. Probably originating along stream banks in the Asian steppes, specifically in Mongolia and northern China, where they still live in underground burrows, they were first scientifically described in Europe by Konrad von Gesner in the famous animal book, he published in the mid-sixteenth century. After this more highly developed, more aggressive rat crossed the Volga and reached Europe in great numbers in the early 1700s, it spread so rapidly that Europeans dubbed it the *Wanderratte* or roving (migratory) rat. Since then it has gone by the names of Norway rat, brown rat, huge mouse, gray rat, wander rat, surmolot, house rat, earth rat, alley rat, water rat, barn rat, sewer rat, dump rat, river rat, and wharf rat, among others. It is known as the Hanoverian rat in England because some historians believe it came there in 1714 on the ship that brought George I from Germany.

The brown rat migrated in large numbers to the New World soon after reaching Europe and quickly spread along the eastern coast of North America, its "*Mayflower*" docking here in about 1755, although it probably didn't reach the west coast until a century later when the forty-niners unsuspectingly carried it with them to California in their search for gold. Ships were largely responsible for distributing the animal, and it wasn't long before they could be found in almost every seaport in the world. Only in the world's warmer areas, such as South Africa, Australia, the South Pacific, the southern United States and South America, did it fail to oust the black rat. But although the black rat is still king of the tropics, by no means has warmer climate completely halted the brown rat's advance. For example, from 1946 to 1952 Norway rats overran one thousand square miles of territory in southwestern Georgia where the black rat had previously been dominant; it had thus in just six years pushed out the black rat in an advance of twenty miles overland in a relatively warm climate.

Brown rats have the widest range of any New World mammal except man, having colonized all of the United States by 1923, when after 148 years the last rat-free state, Montana, fell to the marauders. It has firmly entrenched itself in areas that make up half the world's land surface. Mostly it is associated with man and his buildings, often taking to the fields in rural areas, where it depends on crops for sustenance. One New York City pest-control official has quipped that the brown rat is as domesticated as the dog or cat. Only in dry, unpopulated areas is this rat rare, preferring wet areas. Man's homes, farm dwellings, dumps, slaughterhouses, food-processing plants, sewers, wood piles, compost heaps—any place where adequate food, water, and shelter are available is suitable for the brown rat. But brown rats, too, can survive independently of man when need be. They live apart from humans in Maryland and Carolina coastal marshes and on the shores of Lake Erie, as well as in lonely places like Adak Island in the Aleutians, where they endanger native bird populations by feeding on eggs. On Rat Island in the Aleutians they have the audacity to dig their burrows into the foundations of the bald eagle's cliff nests. Recently, Norway rats were reported nesting in palm trees by the beach in Fort Lauderdale, Florida, where they lived off debris left by people. Such great adaptability, combined with its amazing breeding power, make the brown rat America's number one rat problem.

Laboratory tests have shown that a single pair of rats, either brown or black, could multiply to more than 15,000 descendants in one year if unchecked and to an almost unbelievable 359 million in three years. But many factors limit their growth and they seem to increase just a little faster than man on the average. According to one source, there are an estimated 235 million rats in the United States, a few million more than humans.

Producing a rat population so astonishingly huge requires a staggering amount of sex. Female house rats are theoretically able to bear about sixteen litters a year, as their pups are born in about twenty-two days and they can mate again within forty-eight hours after having delivered them. Rats breed all year round. The female's period of heat lasts six hours, during which time she provocatively wiggles her ears, hops and darts about, a string of sniffing males following her wherever she goes and mating with her as many as five hundred times, which amounts to about one sexual act every forty-three seconds; it seems that the more stimulation a female has the more likely she is to become pregnant.

The rat's promiscuity and phenomenal sexual drive were vividly described by Dr. Stephen C. Frantz, director of the New York State Rodent Control Laboratories, in an account of fifteen to twenty males mating with a single female: "Driven into a sexual frenzy by the female's scent, they mounted her, one after another. The female grew weaker and weaker, but the males couldn't stop. At last, she died of exhaustion, but the males, driven by this fantastic urge beyond their control, continued the orgy, using the dead body."

Male rats are admirably equipped for their sexual role. The male rat has a rough-skinned penis surmounted by a "thorn" for penetrating power, which is needed because he injects sperm directly into the female's uterus, quite aside from any question of whether it gives the female any stimulating pleasure. He mounts the female, often softly squealing, and presses on her flanks with his forelimbs, which provides the stimulation she needs to receive him. She assumes a position called lordosis, with her back arched and abdomen raised high. Once joined, the male thrusts for several seconds and leaps back if there is no ejaculation, sitting on his haunches and licking his penis. Ejaculation usually occurs after six to ten such couplings, the male clinging to the female with his teeth and then falling off to the side of her. In a few moments the same sequence is repeated, this time with fewer couplings before ejaculation; the same process continues, with fewer couplings each time before ejaculation, until the male loses interest or another takes his place. Once the female is impregnated, the male who made her so puts a kind of chastity belt on her. A special male coagulating gland mixes the enzyme vesiculase with the rat's semen and makes the semen set into a soft obstruction blocking the vagina if fertilization occurs. This plug is apparently meant to keep rival sperm away by preventing the female from remating once she is pregnant.

The male rat may use high-pitched sounds beyond the range of human hearing to romance the female, for he certainly does emit such sounds when he has had enough sex. According to Rutgers University researchers Drs. Ronald Barfield and Lynette Geyers, this signal, at a frequency of twenty-two kilohertz, accompanies contented sighlike breathing and "functions as a signal of social withdrawal while the postejaculatory male is recuperating" (a way to say, "Let's rest awhile"). Females, too, use this frequency when they no longer want to be wooed. It is respected by male and female alike, which has led some observers to suggest that broadcasting the twenty-two-kilohertz signal into ratholes could serve as an ultrasonic contraceptive for rats that would limit their population growth.

The female rat's yearly average of up to twenty-two sex acts a day is rivaled by her much larger rodent relative the hamster, which mates up to seventy-five times an hour, well over a mount a minute, when making love. But the gerbil called Shaw's jird is the rodent record holder—one scientist counted no less than 224 mountings during the course of an hour between two of these little creatures—a mounting every fifteen seconds, with a mere three seconds for the sex act and just twelve seconds to rest up for the next one.

Most rodents stay coupled considerably longer than the rat or jird. The agouti, a South American rodent resembling the guinea pig, remains locked together a full twenty-four hours, taking the loving cup, and the Byrnes marsupial mouse or Kowari, not a true mouse, remains in tandem for long periods during which the male thrusts and the female sleeps! Stuart marsupial mice copulate for a continuous twelve hours, but the male finally suffers what one writer has called the ultimate anticlimax. After a mating orgy that lasts about five days every male marsupial mouse that participates, without exception, dies for one reason or another. The cause can be infections, hemorrhages, failed livers, even ulcers. Copulation apparently triggers a fatal overproduction of corticosteroids that brings on the fatal illnesses. It is thought that the fathers being removed from the population makes it easier for the females and their young to survive.

After a gestation period averaging twenty-two days, the female brown or black rat bears six to twelve young, with an average of eight pups and up to twenty pups a litter in rare cases. Generally, the bigger the female rat, the bigger her litter will be. Nevertheless, one study indicates that both the brown rat and the black rat wean only about twenty young per year; that is, only twenty pups survive. Thankfully, for humans, as rat populations increase in size, the rat mortality rate also rises and continues to increase until the population reaches equilibrium and stops growing. During this equilibrium period, the deaths that occur, paired with the number of rats leaving the pack, balances those animals added by reproduction and migrations to the pack. Of the several studies made in this area, one found that for every Norway rat alive at the end of the year, over 16 rats died during the year. In any event, we know that the oldest rat whose age was recorded lived to be five years eight months old in Philadelphia, Pennsylvania, in 1924. This would be equivalent to over 200 years in a human, and there are claims that captive rats can live seven years. But the wild rat usually lives no longer than one year and a study by D. E. Davis, a respected ratologist, put the average life of a wild Norway rat, after weaning, at about six months. Females seem to live longer than males, as is the case with humans, probably because of the greater aggressiveness and wider range of the males, which subject them to more risks of life and limb.

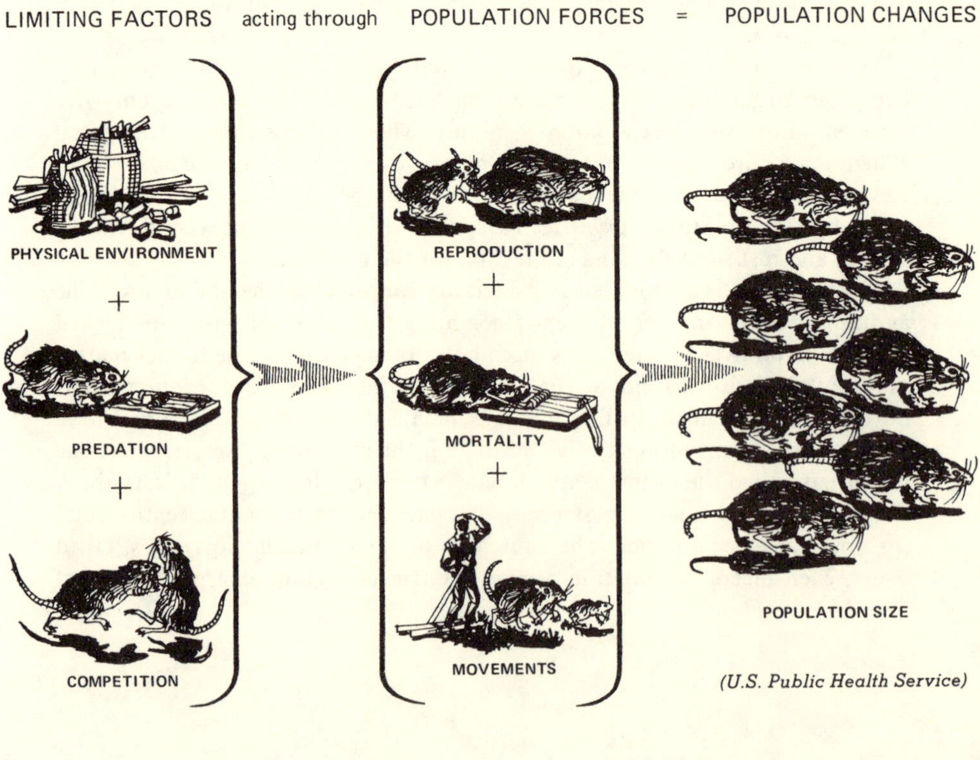

LIMITING FACTORS acting through POPULATION FORCES = POPULATION CHANGES

PHYSICAL ENVIRONMENT
+
PREDATION
+
COMPETITION

REPRODUCTION
+
MORTALITY
+
MOVEMENTS

POPULATION SIZE

*(U.S. Public Health Service)*

Three "limiting factors" determine the balance of the primary forces of reproduction, mortality, and movements in rats. "These three factors are physical environment, predation, and competition," according to Harry D. Pratt and Robert Z. Brown, who studied rodent control for the U.S. Public Health Service. Pratt and Brown point out that the limiting factors effectively control the rate of rat reproduction, mortality, and movement. They go on to define the three factors and their importance. Food, water, shelter, and climate constitute the physical environment, the authors note, while the rat's chief predators are man and various animals, including parasites that cause diseases. Predation, they observe, increases as rat population increases, as does competition, which may be between species (e.g., between black and brown rats), or among members of the same species. These limiting factors are extremely important in rodent control (see chapter 6) and their relation to rat population changes is seen in the accompanying illustration.

The rat population must decrease drastically before the rat's fantastic sex drive is affected. There are always a large number of baby rats or "pups" in every substantial rat pack. Born only 22 days after conception, the young brown or black rat is typically altricial, or blind and helpless. Its pinkish red skin is hairless, folded and wrinkled and its little legs so poorly developed that it can barely move about at first except by wriggling and paddling. The pups hear nothing for about twelve days and usually don't open their eyes until a day or two after that. All their activity is confined to their nests during this time. The nests themselves are generally in safe, quiet places close to food and water. Usually bowl-shaped and some eight inches in diameter, they are made of whatever soft material is available, including cloth, string, rope, paper, excelsior, leaves, and grasses. Sometimes they are in unusual places, such as the nest with live young in it that an investigator found in a space hollowed in ice that had formed around a water pipe on a ship. Roof rats in Florida and Hawaii often build large nests in trees very similar to squirrel nests. Outside, holes and burrows are more often used for nest sites by all sorts of rats. In buildings they use double walls, the space between floor and ceiling, or any protected spaces that enemies can't enter. But the adaptable rodents, especially brown rats, can live happily in sewers, subway tunnels, woodpiles, garbage dumps, rubbish heaps, animal stalls, and even in the depths of suburban compost piles. They are smart enough to use the nests of other animals, too, often while the other animal remains in it. Sometimes they share rabbit holes, which has led to the old canard that rats interbreed with rabbits. In any case, a rat that has no nest will eat her young as soon as she gives birth.

Burrows or tunnels are more often used as nests by the brown rat than the black rat. They are found under or near buildings, walls, earth banks, rubbish, concrete slabs, and similar places. Burrows as long as a city block and five feet deep have been reported. But burrows are usually no longer than three feet and no deeper than one foot, half these distances being the average, and they are often part of an interconnected system enabling a rat to scramble along a two to four-inch-wide passage to a series of nests, food caches, and bolt or escape holes. A dominant male rat may live with a harem of several females in such a burrow system, apparently living in harmony and raising their young communally. Pack size can range from 15 to 220 rats and the larger the pack, the larger and more intricate the burrow system. The rats carry or drag food to food cache sites, where they can eat or rest undisturbed and leave behind feces, food wrappings, and scraps. The bolt or escape holes are used as exits in times of trouble and are often camouflaged with loose vegetation or dirt that an alarmed rat can easily burst through and escape. Building a burrow system is a cooperative effort of many rats and the brown or Norway rat is especially adapted to this work because the ears of this species are small and hairs in the ears keep out dirt. Active burrows can be identified easily. Their entrances will be free of cobwebs and dust, fresh rubmarks will be found on hardpacked soil at the entrances, and there may be fresh fragments of food or freshly dug earth.

The rearing of young rats in the nests and burrows is commonly a communal affair, as would be almost inevitable where there are several females with young in a single nest room, for it seems unlikely that a mother could find and tend only her own offspring among the squirming mass of little pink bodies. This collective raising of the young better insures the security of the following generations, since the other mothers will raise the orphan children of a nursing mother who has been killed.

Warmth is very important to the rat young; since they have no control over their own body temperature they cannot be left unattended for long periods of time. Yet for short periods infant brown rats can survive a body temperature as low as 34° F., while adult brown rats usually die when body temperature drops to 59° F. The rat doesn't gain full control over its own body temperature until it's about 73 days old, when it can maintain a warm temperature even under cold conditions.

Another great danger to the young is disturbance to the nest caused by other rats or outside forces. The very high-strung mother may kill and eat her litter at such times, or she may attempt to move them to another place and kill them in the process. She will also kill her babies if a human handles them and leaves a human scent on them. Usually male rats will not kill nestlings, who are protected by their odor, but sometimes this defense breaks down and the nestlings are slaughtered and eaten. For that matter, rats will devour each other if there is no food available. Take away their food, a veteran exterminator says, and the population will cannibalize itself out of existence.

Whether they are eaten by other rats or fall victim to disease, many infant rats are killed in one way or another before they are weaned. Those that survive are fed by the mother until they are four to five weeks old, although they begin to

A Norway rat by its burrow *(U.S. Department of Agriculture)*

A wood rat *(U.S. Dept. of the Interior, Fish and Wildlife Service)*

This nineteenth-century print accurately depicts a larger Norway rat attacking a black rat.

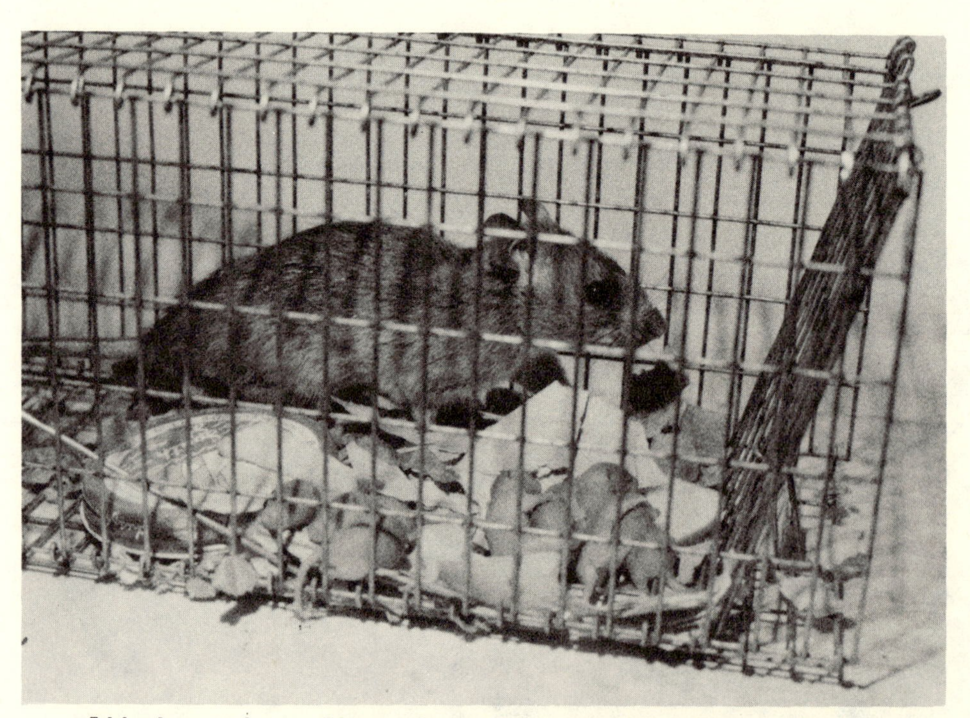

A black or roof rat and her eight newborn pups *(Miami Rodent Control Project)*

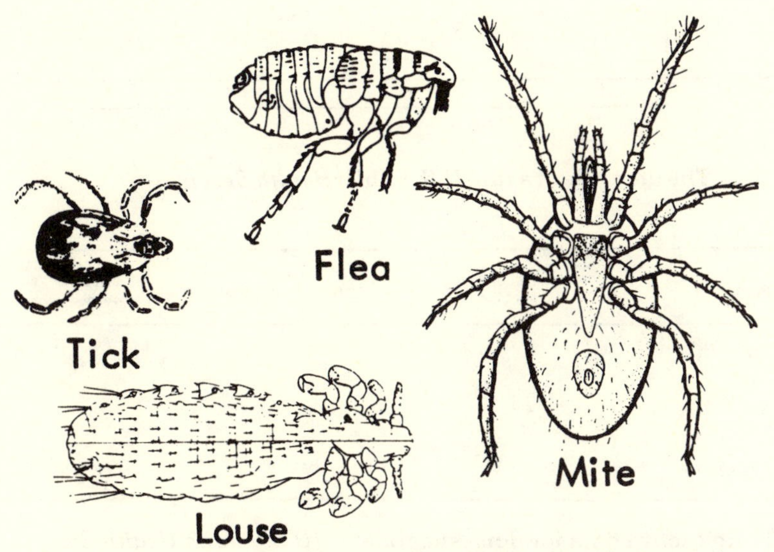

Various insects that infest rats *(U.S. Public Health Service)*

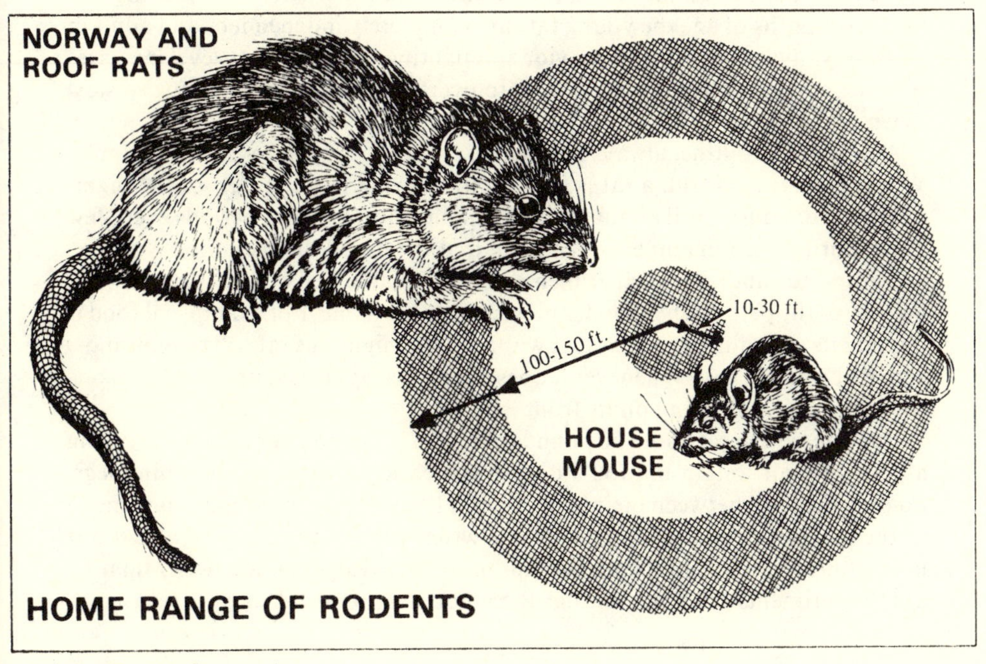

*S. Public Health Service)*

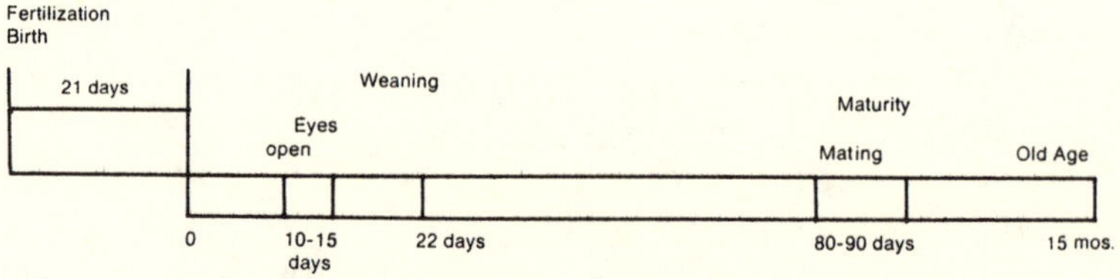

Fertilization
Birth

21 days

Weaning

Eyes open

Maturity

Mating

Old Age

0

10-15 days

22 days

80-90 days

15 mos.

The life cycle of a rat. *(U.S. Public Health Service)*

| Species | Number of Young per Litter | Number of Litters | Number of Young Weaned per Year |
|---|---|---|---|
| Norway Rat | 8-10 | 3-7 | About 20 (18-56) |
| Roof Rat | 4-8 | 3-9 | About 20 (12-47) |
| House Mouse | 4-7 | 3-11 | About 30-35 (23-57) |

Reproductive data for domestic rodents *(U.S. Public Health Service)*

take solid food at three weeks or so of age. The attentive mother even licks each pup's anal region to encourage it to defecate. The pups' first trips from the nest may be accidental ones, taken while they are nursing and clinging to their mother's nipples. They later follow their mother for short distances, increasing the time spent outside the nest until finally they accompany her as she goes about all her activities. If they get lost or encounter trouble, they utter ultrasonic squeaks that the mother hears and responds to. The mother does not consciously try to teach her babies; they learn by imitation and experience. On reaching two to three months of age the young rats are completely independent and adult in every way, including sexual behavior and fighting. Their activity level increases for nine months, when old age begins to overtake them, and they start to slow down.

A rat's daily routine, always a frenetic rat race, is related to the food supply. Where food is plentiful, a rat is more active during the first part of the night, ranging from dusk until about midnight; it has shorter active periods every few hours during the remainder of the night and the day. But when food is scarce, according to laboratory experiments, the adaptable rat will shift its peak of activity to the period when food is most readily available. For example, if food is put into its cage during the day and withdrawn at night, the rat will become most active during the day. Whenever it does sleep, it sleeps rolled up, resting on one side with its head resting on its front legs.

Rats are said to eat about 50 pounds of food a year and to be capable of eating a third of their weight in food a day. When eating they sit on their hind legs, holding the food between their front paws. They will eat almost anything, much to the misfortune of man, although one writer reports that they *won't* eat our ubiquitous refined flour, being perhaps more knowledgeable and wiser than we are. Experiments have shown that the amount of food taken by a rat is controlled by its pituitary gland and if the gland is injured it will eat constantly, commonly doubling its weight in a mere two months. A second control center next to the pituitary stimulates the rat to eat or not to eat, acting like a thermostat, and if it is damaged the rat will refuse food and grow emaciated. Other experiments have shown that rats will eat sweets much like a person let loose in a candy store with access to candy of all kinds. At first they will eat the sweetest candies and eat to excess, but they will quickly show a preference for intermediate-strength sugar solutions of about 10 percent sucrose, preferring the moderately sweet.

Rats can survive up to 14 days without food, their weight usually halved and body tissue damaged after two weeks of starvation. If they are allowed an abundance of food, and can eat as much as they want, they will be more active and learn more rapidly than if only fed adequately, according to experiments made at Cambridge University.

Grain-producing regions in Asia were the original home of the rat and grains are still a staple of its diet, but rats often show a preference for animal matter, such as meat, fish, and eggs. They have been known to kill and eat mice, chickens, ducks, geese, and birds and will even gnaw on live young lambs and piglets, or bite large pieces from adult pigs. It is a documented fact that several elephants in Germany's Tierpark Zoo died when rats gnawed at their oil-rich toenails. A Zurich zoo reported the death of ostriches there because rats had inflicted wounds that wouldn't heal. Neither is there any doubt that rats have eaten flesh from helpless people and infants (see chapter 1).

Excellent fishermen, rats have a particular behavior pattern for searching the water for food; it is known as the "searching grasp," a sievelike motion they make with their feet in the current. They like fish enough to make nightly mile-and-a-half-long excursions onto the delta of one north German village to raid the eel nets. As noted, their fondness for birds' eggs has led to the near-extermination of bird colonies on several islands, and they have been reported to dive for fish in small ponds and zoo aquaria.

Rats will even eat manure and drink human urine if there are no other food sources. Being hoarders, they cover their food with whatever material is available as soon as they can. According to one study, their favorite foods include grain and grain products such as oatmeal, potatoes, meats, and cooked eggs, while they seem to show little interest in such foods as raw beets, peaches, onions, celery, cauliflower, and green peppers. But their choice of food is determined largely by their environment. For instance, they normally don't prefer citrus fruits, but in areas of Florida where there is little other food to choose from, black rats are a serious pest in the orange groves. In cities they feed primarily on garbage, which many exterminators call "rat food," eating about 1.3 ounces of it a day. Apparently the garbage is a well-balanced diet, for controlled experiments have shown that garbage-eating city rats are considerably bigger than their corn-fed country cousins. Rats do seem to be able to balance their diets. Rats used by Dr. Curt P. Richter of Johns Hopkins in a five-year study were given free access to separate bins of protein, carbohydrates, fats, vitamins, and minerals. They soon selected each in healthful proportions.

In their search for food, rats range over a surprisingly small area. The size of a rat's home range of course depends on local conditions, including primarily the presence of food, water, and hiding places. But studies have shown that Baltimore brown rats seldom stray more than 100 to 150 feet from the nest and that Hawaiian black rats tend to remain within a radius of 100 feet. Other studies have established home ranges of a mere two or three feet—this in the case of rats living under open garbage cans near a leaky water spigot. Rats will migrate seasonally, or from flooded or burned fields, and released captured rats have reportedly traveled as far as four miles trying to find a suitable home, but these are special cases.

In contrast to nibbling mice, rats are steady feeders. They all seem to have a need for privacy when eating and like to carry their food to hiding places, except when it is small enough to bolt down on the spot. Unlike many rodents, rats also have a strong need for water and drink ½–1 ounce of it a day, although they can survive on much less. In order to get water they will go to such lengths as licking dew from the grass or gnawing through lead water pipes, causing much damage.

The rat gnaws fully half of the time it is awake and it absolutely must if it wants to live very long. This is not a matter of eating. "Rat," like "rodent," means "gnawing animal" and few other creatures have more effective cutting tools than the eponymous rat's front or incisor teeth, which can exert an incredible pressure of *24,000 pounds per square inch* and can cut through paper, rags, wood, bone, insulation of all kinds, asbestos, brick, cinder blocks, four-inch thick concrete, aluminum, and even half-inch sheet metal. These incisor teeth appear eight days or so after birth and the rat begins gnawing of necessity almost at once. Definitely "of necessity," for average growth of a rat's lower incisors is 5¾ inches a year and average growth of its upper incisors is 4½ inches a year. Over a three-year life span this amounts to a total growth of over a foot for the lower incisors and almost a foot and a half for the upper incisors. It is obviously a matter of keep gnawing or die. Any rat who didn't keep gnawing and grinding down its incisors would eventually have permanently locked jaws that would prevent it from eating, or a lower incisor might push up so high through the space created by a lost upper tooth that it would pierce the rat's brain and kill it.

Gnaw the rat must and it does—with a vengeance, stopping only to find something else to gnaw upon. No myths are those previously recounted stories about the rat's fantastic destructive forays, its vandalizing of everything from poultry, livestock, crops, and warehoused foods, to electrical wiring. Such destruction is not all wanton, for the rat must gnaw. The black rat is only a slightly better gnawer than the brown. Recently it has been suggested that rats grind their teeth *all the time* and that these are the noises largely used in their echo-location. The rebounding echoes of the rat's grinding teeth, it seems, enable it to choose paths without obstacles, much like the bat's sonar. Another researcher has theorized that rats grind their teeth at all times except when they are disturbed or frightened—so pleasant is the grinding process to the rodent.

Using their tails for balance, rats are superior climbers, have been seen scurrying along telephone wires and have indeed been trained to walk tight-ropes. Their excellent sense of balance can be tested by tossing one in the air; even when only a few days old a rat will invariably land on its feet. Rats have been known to fall five stories to the ground and scurry off as if nothing had happened, and they often enter buildings by dropping through open skylights. The black or roof rat is a better climber than the brown rat, but both can scale a sheer brick wall as if it had stairs and cross city streets by walking telephone wires. Rats can climb any vertical wall on which they obtain a claw hold and they manage to ascend some perfectly smooth walls by using a pipe, a corner, or something else on which they can brace their backs. They have been found climbing up buildings inside 1½-inch diameter pipes attached to the structures and can cross sheet-metal flashing by catching the top edge with their front claws and swinging across overhand. They only perform these acrobatics when driven to it by hunger or lack of shelter, but are fully capable of such performances. They can also jump upwards two feet and manage a leap of three upward feet with a running start. Horizontally, rats have jumped four feet. Jumping out and down from a height of about fifteen feet they can cover a horizontal distance of eight feet. Add to this the facts that they have a reach of thirteen inches and can squeeze through a hole no larger than a quarter dollar. They make superb athletes indeed.

A New York police launch once clocked three Norway rats swimming in rough water across the Hudson River at 2 hours 50 minutes. All rats are good swimmers, and the brown Norway rat is semiaquatic by habit, frequently living by streams, in marshy areas, and in sewers. But while brown rats are known as sewer rats, black rats will take their place and thrive in the sewers if the opportunity arises. There are stories of rats surviving after being flushed down toilets and numerous cases have been reported of sewer rats swimming through sewer lines up into toilets many stories above the street. Rats in large cities use the sewer lines as regular highways. They can swim as far as half a mile in the open sea and tread water for three days, have swum for hours when placed in tanks and can dive and swim underwater for as long as 30 seconds at a time, diving 100 feet down. Rats have been found swimming three days after ships went down in midocean. In an early study for the U.S. Fish and Wildlife Service, Charles Cottam had this to say of the brown rat's familiarity with water:

An example of the Norway rat's adaptability was recently brought to my attention at the U.S. Fish Hatchery Station in Leetown, West Virginia. Despite the fact that several buildings and food-storage bins are kept reasonably free of rats from 50 to 200 yards distant, a surprisingly large colony was found living in earthern banks adjacent to the fish ponds and stream courses where concentrations of fish are kept. The spawn, fingerlings, and adult fish—primarily of species or varieties of trout and bass—are fed horse meat daily as well as other prepared foods that are equally acceptable to the rat colony. Undoubtedly, bits of food, accidentally dropped on the bank in feeding operations, accustom the rats to that type of diet. It is obvious that an inadequate food supply occurs on the land, and the rats have learned to obtain food from the water. In their competition for food, the young and small fish at all hatcheries concentrate in enormous numbers when food is thrown into the water. Consequently, food is available for a very short period only. The rats as well as the fish have learned this. Therefore, at this hatchery these normally nocturnal rats have synchronized their feeding time to correspond with the feeding of the fish. When food is thrown into the water, the rats have no hesitancy in competing for it. They swim well and rapidly. After the food in the water is consumed, they then seek tidbits inadvertently dropped on the bank. Apparently low temperature is no serious deterrent to the rat's aquatic activity. On the day of my visit . . . the temperature was considerably below freezing, and an unpleasant sleet and snow storm was in progress; even so, a dozen rats were observed in the water, and part of their runway extended through small areas of the stream. While omnivorous rats avidly consume all kinds of fish food, it was a surprise to find that they are not at all averse to preying on young fish. On a number of occasions the rats have been observed catching the fingerling fish and taking them into their ditchbank runways or burrows.

Rats love water and delight in bathing in it. There is really no truth in the saying *a dirty rat,* as we have noted, for the rat is only soiled by man's dirt; he is naturally quite a clean animal. One researcher meticulously observed a wild rat on rising: "Eyes open. Rises and stretches. Licks hands; washes face; washes behind ears and continues to lick hands at intervals. Licks fur of back, flanks, abdomen. Licks hind toes, scratches with them; licks scratches . . . scratches flank and belly. . . . Licks genitals. . . . Licks tail while held in hands. . . . Licks hind legs held in hands. . . ." This washing and grooming is a daily ritual and unlike the cat, the rat uses both paws in performing it.

Without keen senses no rat would long survive and many of a rat's senses are sharp from early in life. Vision is probably the least developed of rat senses compared to humans. Despite their bright eyes, rats are very near-sighted and one can slowly bring his face to within a few inches of a rat without provoking a reaction. Rats do, however, have specialized nocturnal vision, which enables them to recognize motion up to 30 feet away and to identify shapes in dim light. They are color-blind; color appears to them in various shades of gray, as it does to color-blind humans. Rats are relatively insensitive to red light and can be closely observed in a darkened room if a reddish light is used.

The rat has an excellent sense of smell, which enables it to both find food and follow rodent body odor, especially the odor of female rats. Its sense of touch is especially important because the rat so often functions in the dark. Besides the normal ability to feel with their bodies, rats have highly sensitive vibrissae or guard hairs, whiskers that are actually longer hairs scattered throughout the rodent's fur and are more sensitive than the shorter fur covering the body. This accounts for their preference for running along walls or between things with which their guard hairs can keep in contact.

An acute sense of hearing is also a necessity for the nocturnal rat, which can easily recognize noises and locate them to within a half foot. One story, yet unaccepted by scientists, says that rats like music so much that they have clicked their teeth to applaud it. Experiments have shown that they are practically deaf to low notes but highly receptive to shrill ones, hearing notes that are too high in pitch for the human ear. Rats produce two types of sounds: biosonics, which are duplications of natural sounds, and ultrasonics, high frequency sounds beyond the range of human hearing. When alarmed, baby rats in the nest call their mother with ultrasonic cries. Adult rats display twitching movements of their long whiskers and large delicate ears when they hear high-pitched sounds. "If you go into a room where rats are kept and talk in an ordinary voice you can see the rats wincing in unison every time you come to a sibilant," British researcher R. J. Pumphrey wrote in a 1950 study. But ultrasound has not yet succeeded in killing rats or driving them from buildings, as many inventors have hoped. The best this expensive eradication method has done thus far is to reduce the number of rodents in a location, principally because the method is directional and produces "sound shadows" where rodents are not affected.

Rat taste in foods parallels that of man, one reason the creature has thrived so long, but the rat has a more highly developed sense of taste. One researcher showed that the brown rat could discriminate between plain bait and the same bait containing as little as two parts per million of an estrogen, concluding that "this unusual ability to detect extremely minute quantities of bitter, toxic or unpleasant substances" is what makes the rat so resistant to poison.

Rats seem happy only when they are busy, an extreme example being the pack rat, which is obsessed with the desire to move things. All of them are collectors like the pack rat, stealing objects and carrying them off to their nests (see chapter 2). Definitely true is the old story that a cornered rat will fight to the death; a desperate rat will attack cat, dog, man—anything that blocks its escape. Rats have been known to combine forces to kill cats and dogs when cornered. It is true, too, that the rat's entire life is a rat race; rats seem to epitomize frenetic action and have been recorded doing over twenty miles a day on treadmills in their cages, though their top speed is only six miles an hour.

A rare tropical tree is actually pollinated by rats, which is one of the few rat contributions to the world outside of the laboratory. It has been observed that rats can develop certain behaviors beyond their innate "behavior inventory"; such behavior, always confined to a local population, can be passed on to younger members of a pack and has been called "tradition," an unusual phenomenon among lower mammals. Rats, for example, have learned to fish by extending their forepaws into the water, and there are those who swear they can steal food by dipping their tails into bottles too small to admit their snouts. One writer told of observing rats fishing with their pink, naked tails on the edge of a coral reef in the Trobriand Islands: "Rat after rat dangled his tail in the water. Suddenly, one rat gave a violent leap of about a yard and, as he landed, I saw a crab clinging to his tail. Turning around, the rat grabbed the crab and devoured it, then returned to his perch. Meanwhile, the other rats were repeating the same performance." The French scientist Dr. A. C. Calmette told of a scientist finding sixteen large live crabs with their legs cut off in a large rat burrow on one of the Channel Islands. "Investigation showed that the rats were in the habit of making crabbing expeditions at low tide. To immobilize their victims and render them harmless the rats amputated them as soon as captured. All the crabs found were still living and in good condition. Whether the wily rodents kept their prisoners fed or not is unknown."

Maze tests given at the Museum of Natural History's Animal Behavior Laboratory proved the rat to be an adaptable, curious, intelligent animal. Rats in other tests learned to select one letter of the alphabet from a row of mixed ones, to pick a given inkblot from several different ones, and even to identify an advertisement for ice cream and then select it from a dozen assorted ads. Although some researchers argue that lab studies demonstrating the ingenuity of rats really reflect the ingenuity of experimenters in making rats appear intelligent, the rat's problem-solving abilities have been confirmed by recent studies of rats under natural conditions. Yet there are limits to a rat's intelligence. As Isaac Asimov has pointed out, a rat can associate mission with assignment for only about twenty-five seconds. The rat seems to be stumped if the delay between pressing a lever and receiving food stretches beyond that time.

Love seems to raise a rat's I.Q. Some twenty-five years ago, experimenters tested a large number of infant rats by leaving a third of them in their nests unhandled, gently handling a second group of equal number and subjecting a third group to both handling and mild electric shocks. All three groups were given adequate warmth, food, water, and clean nests. Surprisingly the group handled but manipulated by electric shock fared much better than the group that had been left entirely alone. The unhandled group cowered in corners, were generally timid, urinated frequently, and could not bear temperature changes or other hardships nearly as well as the other two groups. The group handled with love and care ranked first in intelligence, friendliness, and all other qualities. More recent experiments have confirmed these findings, leading some psychiatrists to see a parallel between the untended rats and human children left in foundling homes. The value of handling rats has led many research labs using large colonies of rats to employ professional "gentlers" whose job it is to regularly handle the rats so that they remain gentle and unneurotic.

Rats appear to respond in human ways to their surroundings as well as their treatment. In a 1954 experiment at the Center for Advanced Studies in Behavioral Science of the National Institute of Mental Health (NIMH) twenty male and twenty female rats were penned in a very large enclosure under what seemed ideal conditions with plenty of food and water. But the rats found the area too crowded and there was a breakdown in normal behavior. Mothers failed to care properly for their young, courtship was less frequent, the rats were irritable and jumpy, many living by themselves, fighting senselessly and often to the death. The rats had responded in humanlike ways to the overcrowding and a population that should have grown to thousands remained steady at about one hundred and fifty.

Warfare among rats is common when strangers enter a colony's territory. When this happens the whole colony is alerted by what is called "mood transmission." Black rats transmit this alarm with shrill calls, while brown rats use only body language. In one observed case "with eyes bulging from their sockets, their hair standing on end, the brown rats went out on a rat hunt," the hunters so keyed up that they often bumped into each other and fought among themselves. Pursued rats are either systematically torn apart or die of shock (which also sometimes happens when a human picks up a wild rat). They do not fight back. Even those suffering no physical injury usually die within a few days, these victims showing enlargement of the adrenal glands, which makes some scientists believe that their deaths are caused by an overreaction to stress.

An experimenter who put seventeen strange rats into a large natural enclosure found that the first two to pair off terrorized all the other rats and prevented them from bonding. Over a three-week period all the others were killed by the mated pair, sometimes from a direct bite on the neck which pierced the carotid artery, from infected wounds, or from sheer nervous exhaustion.

Rat fights can be terrifying to watch, involving evasion, circling, squealing, chattering of teeth, tail rattling, freezing or catalepsy, aggressive grooming of each other, clinching, kicking, pushing, leaping, vicious biting, and even boxing. When boxing, both rats rear up on their hind legs and hit each other with their front paws, often for as long as a fifteen-round human boxing match. The victor usually humiliates its defeated, demoralized opponent by crawling head first under its body.

Competition among rats is most intense between the brown and black species, and between established members of a colony and invaders, but social order within any colony of the same species is also determined largely by fighting. Dominant rats, those who win their fights, are the ones that live nearest the food supply in a colony and the farther a rat is from the food the lower his social rank. Furthermore, those rats living farther away from the food have to run a gauntlet of rats to get their food and are often attacked by them. The lower status is passed on to future generations as well, for the weaker adults get less food and their young, less numerous, do not grow as fast or as big as the progeny of dominant rats. Being weaker and more often wounded, the lower rats are of course also those most frequently killed by predators such as rats, dogs, and humans.

Konrad Lorenz believes that constant warfare between rat clans increases their ability to fight and "probably has put a premium on the most highly populated families," so that their fighting nature ultimately is not detrimental to the rats' survival in this world.

Those tales of great rodent nations marching into battle are pure myth, ratologists say, but no one really knows the rat's full capabilities. Surely there are many stories of rat cooperation and even compassion. A good example is the famous *Rattenkönig* or "rat king." Young rats close to one another in the nest sometimes get their tails entangled and become a living Gordian knot glued together by dirt-encrusted wounds and the like. When they try to pull apart the tails are pulled tight, and the knots strengthen, knitting the rats together. As many as 32 rats are trapped in these knots and have died as a result of being unable to forage for themselves. However, they are often unselfishly fed for life by other family members.

There is no doubt that rat kings exist; sixty or so have been reported in Europe since 1564 and about forty (most of them found alive) have been authenticated, the latest in 1963. Rat kings have frequently been preserved, painted, and photographed, and in 1774 a 16-rat king was examined by a Leipzig court in connection with a charge that a miller's apprentice had cheated his master by stealing the king from him and pocketing a tidy sum by exhibiting it.

The name rat king may come from the old superstition that an aged wise rat sat on the entangled tails of rats and was treated as royalty by the pack. But it could just as well derive from an early belief that the animals entangled were one organism, a supreme rat with many bodies. Rat kings range in composition from 3–32 rats, with most consisting of 5–10 animals, and are apparently found only among the long and less pliable-tailed black rat species, although a few verified rodent kings of squirrels and several unverified mice kings have been reported. Brown rat kings have been induced in the laboratory.

Rat kings fabricated by tying the tails of live rats together look nothing like real kings, but rat kings have been created in the laboratory by gluing the tails of rats together; this causes the rats to become so entangled while trying to extricate themselves that a true knot is formed. Yet no zoologist has been able to prove exactly how rat kings are formed in nature. It is possible that the tails become entangled when the rats huddle together facing outward for warmth and security, urine and feces from those in the upper circle falling onto the entwined mass of tails. Other possibilities are that the tails might become entangled while the males are wildly fighting for females, or during mass grooming, or in the nest shortly after birth, or after the tails of a number of rats come in contact with some sticky substance. It may even be that the "verminous vermicelli" are formed in several ways. The rat king remains as much a mystery to nuclear-age scientists as it was to medieval peasants.

Even more mysterious are the many tales, all scientifically unauthenticated, of rats and other rodents falling from the sky in great numbers. As mentioned, the *Wall Street Journal* carried a story about such a phenomenon in recent years. Rat showers have been reported since medieval times, especially in Arctic regions. Olaus Magnus, Erasmus Francisci, and other early naturalists cited contemporary accounts of rodent showers, which were said to occur during wind storms, or following rain and dense fog. Francisci said that hordes of rats had been found on snowcovered mountaintops after wind storms when there were no tracks leading up the mountain and the ground was so frozen that the rats could not have emerged from the ice-blocked burrows there. The most common of such rodent showers are those of lemmings, which the Eskimos believe fall through space from another galaxy.

Several stories have it that rats caught by their tails in traps escaped when comrades came to their aid and bit off their tails. While this has certainly happened, some observers say that it is no example of cooperation or compassion but an illustration of the fact that rats will aggressively attack any other rat showing abnormal behavior. More difficult to explain are the stories of adult rats pushing food down off a picnic table for smaller rats who could not climb it, or stories of rats fording a river, each swimming and holding on to the tail of the rat in front of it.

A century ago, in the journal *Nature,* British biologist T. W. Kirk told of a brown rat and a helper dragging a four-inch biscuit through bars two inches apart:

> After several unsuccessful attempts the lone rat left it, and in about five minutes returned with another rat, rather smaller than himself. He then came through the bars, and, pushing his nose under the biscuit, gradually tipped it on edge, rat number two pulling vigorously from the other side; by this means they finally succeeded in getting a four-inch biscuit through a two-inch aperture.... It was evident that the first rat saw that to get the biscuit through the bars it was necessary that it should be on its edge, and not being able to tip

it and pull it at the same time, he gained the assistance of a friend. The concerted effort shows also that rats must have some wonderfully facile means of communicating ideas.

There have been many, many reports of "compassionate" normal rats leading blind rats, each holding one end of a twig or straw. It would be easy to mistake two normal rats carrying away a small bone or straw, each wanting it for his own, as a normal rat leading a blind one, but in at least one case a scientist has confirmed the story. British naturalist Eric Simms and a companion came across two rats walking along, one holding on to the tail of the other. After his companion shot the two rats Simms examined them and found that the rat holding the tail was blind in both eyes.

The most persistent story about rat cooperation concerns two rats transporting an egg in an unusual way. This tale's origin is lost in history, but it has been told scores of times (there are seven or so accounts of it in Maurice Burton's *Animal Legends* alone), and it may well be true despite the objections of several zoologists. In brief, the rat is said to transport the egg by lying on its back and holding the egg under its chin with its forepaws while another rat drags it along by the tail. La Fontaine wrote a fable about a rat carrying an egg in this manner

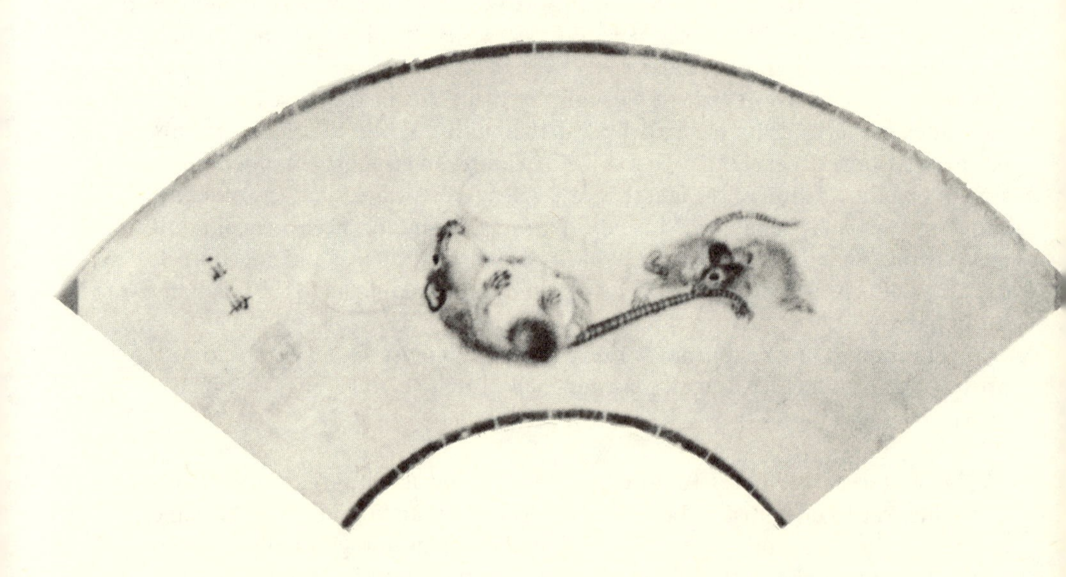

Two rats cooperating in stealing an egg are shown in this detail from an eighteenth-century Japanese fan painting by Satake Eikai. *(British Museum Japanese Painting Collection)*

and the subject fascinated the early Japanese, as a number of old prints show. There is no doubt that rats can be taught such behavior. It has also been suggested that rats roll the eggs along the ground with their forepaws; that they form a line and hand the eggs from one to another; and that rats are carried by other rats while holding eggs. One author claims that rats can carry eggs up and downstairs, passing the eggs to each other with their front paws when going downstairs and with their back legs when going up. Rats definitely can carry an egg by biting two small holes in its shell and hooking its teeth into the holes, but many unmarked whole eggs have been found in rat burrows. Whatever the answer, here is one of the first recorded accounts of a rat being dragged along by another rat, from a letter of British naturalist, Henry Moses, to the *Zoologist* in 1865:

> The rector of a parish in Westmoreland assured me that he had witnessed this feat. Having lost many eggs belonging to a laying hen, he was induced to watch to discover the thief. One morning, soon after the cackling bird had given warning that she had deposited an egg, he observed two rats come out of a hole in the henhouse and proceed direct to the nest. One of the rats then laid down on its side while the other rat rolled the egg so near it that it could embrace it with its feet. Having now obtained a secure hold of its egg, its companion dragged the first rat holding the egg into the hole by its tail.

Despite myriad behavioral studies of the rat, its behavior still seems mysterious and unfathomable to the most industrious rat writer or reader. No matter how carefully and persistently learned men study this creature of the night, it always seems to be just beyond our grasp, eluding human understanding as it does human traps. As a final example, here is a tale which might make one believe that the myopic rat can read! Only this ability or the possession of some sixth sense, unfathomable to man, could explain what happened in 1970 when the French relocated the renowned Parisian marketplace of Les Halles several miles away to Runges. Over two million rats made the move as well, all of them traveling underground. This would not be at all unusual if the rats *followed* the provisioners to the new marketplace, or came a few days, weeks, or months later, for rats do have that acute sense of smell which leads them to food. But the Les Halles rats—all two million of them—were there waiting for the first trucks to arrive at Runges long before a shelf was restocked! How did they learn of the move in advance? Did they subscribe to *Le Monde*? No one knows, or in the words of one scientist, "no satisfactory theory has been put forth to explain the phenomenon."

# V

# RAT KILLERS

The rat moved restlessly in the walls. The dull scramble became more pronounced but suddenly ceased when at dusk a young woman entered the room and bent over her baby's crib. The rat was watching with beady black eyes from a hole no bigger than an inch in diameter near the radiator.

Gently, the woman kissed the child on her warm, tender forehead and propped a bottle in her little mouth. "Goodnight baby," she said, and she turned from the bedroom, leaving the door slightly ajar. Dog at her feet, she sat out of sight of the rat in the adjoining room.

The rat peered from the hole for only seconds as the baby sucked on the bottle hungrily, her chest heaving rhythmic sighs of satisfaction. Milk trickled from the too-large hole in the nipple, dribbled off the infant's chin, down her jaw, and into the folds of her throat.

All of the rat's senses, its entire being, drove it toward the warm tender flesh and the milk. The stained white belly pressed low against the floor. The greasy gray fur, black with filth and vermin from its slick runways, helped it slide the pliant body through the narrow hole, first the long sensitive whiskers emerging, the prominent teeth sharp and white from constant gnawing, finally the scaly pink tail.

The baby cooed and sighed. Stealthily, the rat approached the sounds and smells, the unmenacing little body four times as big but no match for it in speed and strength and cunning. Once at the base of the crib, sensing no danger, the rat leapt up and wriggled through the bars.

Instantly, it was between the infant's folded legs smelling at the diaper: whiskers, tail, and paws scraping against skin; fleas from it finding warmth in the bedding as it hopped up on the child's bare chest. Almost in the same motion it was at the beads of milk like a necklace around the tiny throat.

Now the infant from deep inside herself, perhaps from out of her ancestral past, let loose the bottle from her lips and wailed. A shrill cry that became a shriek alerting mother and dog. Sharp cries of pain when the startled rat struck back at the little body's sudden fearful rejection, biting at throat and cheek and lips, all in what seemed one darting motion.

The dog was first into the room and the rat instantly saw him, sensed the dog blocking its path of escape on one side and the woman, broom in hand, blocking the doorway. It leapt toward the woman, and when she raised her broom, darted up the long broom handle, slashing her fingers and tearing the thumbnail off her right hand. But the dog was upon it just as quickly, snarling fiercely as he cornered the rat. The dog—a terrier, a line bred for ratting—struck out, snapping at its neck but missing, and the cornered rat sprang up catching hold of his nose and clinging tight. The dog twisted and writhed trying to shake the furry thing loose and at last succeeded, smashing his quarry against the wall and finally crunching the rat in his strong jaws.

The rat, ripped apart now on the bloodstained floor, had seriously wounded infant, woman, and dog, but the dog had luckily done its job well and saved the child's life. Dogs have often performed this heroically. Dogs have, in fact, always been better rat killers than most animals throughout history and are certainly more courageous when facing rats than are cats. Judging by animal intelligence tests, if their accuracy can be accepted, the wily rat is about as intelligent as the dog, perhaps a bit more intelligent. The rat will combine with others to achieve goals as a dog will, for example, and seems capable of rudimentary planning. On balance, the rat outsmarts the dog more than the dog gets the advantage of him (all those fairy tales and cartoons of rats and mice outwitting cats and dogs are basically true), but the dog does prove a good ratcatcher in certain situations. Unlike mice, which have been known to drop dead when subjected to sudden noises such as the snap of a moustrap, rats are not timid souls and will attack the largest of dogs if cornered. Often they are able to make their escape from such a startled dog, but not from the courageous dogs trained to be ratters, who prevent all but the most foolish rats from leaving their burrows and running rampant through a farm or building.

Dogs weren't valued as ratters in medieval times, when there were more human "dog killers" than "rat killers" and forty thousand dogs were exterminated as plague carriers during the Great Plague in London. But American and European newspapers of the last century are filled with owners' proud accounts of how their dogs slaughtered hundreds of rats at a time in rat-killing contests where scores of rats were emptied into a small pit and a dog was set loose ripping and slashing through the squealing mass of creatures. Some descriptions of this bloody spectator "sport" are enough to make one feel sorry for rats, but it was immensely popular among Englishmen, Americans, and especially Frenchmen, who held fights along the present Rue Maitre Albert when it bore its ancient name carved in stone: *Rue des Rats.* Ratting became the main indoor blood sport after British cockfighting was outlawed in 1849, and was enjoyed by rich and poor alike. "Rat-killing Legers," in which dogs were matched to see which could kill the most rats in the shortest time, were still being held at public houses throughout England early in the 1900s. Signs in the streets reading "Rats Bought and Sold" or "Rats Wanted" were common, and professional ratcatchers made a good living ridding houses of rats for low rates, provided they could take the rats live and sell them. Country rats commanded the best prices for ratting matches because they were comparatively free of disease, but even sewer rats were used in a pinch. A sporting pub might buy as many as two thousand rats a week, and a superior country rat specimen brought as much as a shilling. Before the rats were loosed into the pit and a dog set upon them, the spectators would blow on them, raising their fur and sending them into a frenzy.

During the 1870s and 1880s, American newspapers were full of stories about prize rat-killing dogs, or, where rat fights were illegal, about police raids on private homes, pubs, or stables where scores of people were arrested and great numbers of rats and dogs confiscated. Following is an 1878 account from, of all places, *The New York Times* sports pages:

## A Dog's Marvelous Feat

A large number of sporting men assembled last night in a well-known up-town sporting resort to witness a ratting match for $100 a side between the imported black-and-tan bull terrier Harry, and Flora, a white bull-and-terrier slut. The conditions of the match were that each dog should be put in the pit with 50 rats, and the dog killing the rats in the shortest time to be declared the winner. A number of wagers from $5 to $50 were made on the result, Harry being slightly the favorite. After the usual preliminaries had been completed, 50 average-sized rats were dropped from a bag into the pit. Flora, who was handled by her owner, was then set to work upon her task. From the start she killed with astonishing rapidity, and in less than a minute 10 rats lay dead in the pit. She continued on her task without stopping, and in two minutes she had killed 30 rats, and, although bitten several times by the vermin, she accomplished her task of killing the 50 rats in the wonderful time of 4 minutes 15 seconds. After a recess of about 5 minutes the imported dog Harry was brought into the room. He appeared too fleshy for work, and bets were made that he would not beat Flora's time. The rats were put into the pit, and the dog immediately after. He set to work in a very resolute manner, but after killing a few he seemed to hesitate several times; he however, finally, after much urging by his owner, killed all his rats, the time being 9 minutes, 30 seconds. Flora was declared the winner. The owner of the imported dog said that he attributed his dog's defeat to its being out of condition. He was willing to back him against Flora for any amount to kill 100 rats. It is probable that another match will be made.

For the record, the most vaunted rat killer among dogs was a bullterrier bitch named Jenny Lind (after the "Swedish Nightingale"), who killed 500 rats in 1½ hours at an arena called "The Beehive" in Liverpool, England in 1853. The British bullterrier, "Jocko," killed 1000 rats in 1 hour 40 minutes, but the feat was performed over a period of ten weeks in 1852, Jocko loosed upon 100 rats at a time, which wasn't exactly cricket. One benefit of the sickening ratting contests was the breeding of albino rats (used because the blood showed better on them), which were eventually put to use in scientific laboratories.

Terriers are considered the greatest rat hunters among dogs, and are still used extensively in special situations. When rats threatened the palm-seed industry on Australia's Lord Howe Island, for example, rats were smoked out of hollow trees and chased with terriers if they attempted to escape by ground. But like cats, dogs generally do little more than keep rats out of sight, in places where they can't be reached, which limits their spread to new areas. As a matter of fact, one of the most common foods rats eat in residential areas is the food that's put out for dogs, and in one study the largest rat caught was trapped under a doghouse.

Dogs may be better at the job, but no animal save man has a better reputation as a ratcatcher, merited or not, than the common cat. In early Egypt the cat was trained to guard grain warehouses along the Nile against the ravages of rats. These highly regarded animals, and many other Egyptian cats for that matter, were embalmed and buried with great ceremony when they died, a number of embalmed rats buried with them for their dining pleasure in the afterworld.

As soon as the cat was imported to Rome from Egypt in about 150 B.C., it became almost an arm of the municipal sanitation system. But the feline grew to be so venerated, loved, and well fed in Rome that today, after two thousand years, it seems to have lost most of its ratcatching instincts. The over 250,000 feline vagrants in Roman ruins, such as the Forum and Pantheon, have an estimated 30 million Roman rats they could feed upon, but so accustomed are they to food handouts that they rarely catch rats anymore. In fact, with the decline of the old *gatara,* or "cat ladies," who have traditionally fed them, Rome's cats are reportedly joining in the human flight to the suburbs.

Medieval legend tells the story of a cat who rolls in red mud so that she looks like she is covered in blood. Then Tabby throws herself on the ground, holds her breath, and waits for a rat to approach. The rat, observing that she is not breathing and that she seems covered with blood with her tongue hanging out, thinks that she is dead and sits on her. Tabby then grabs the rat and gobbles it up.

Chaucer has lines about the cat as a vaunted, driven, rodent catcher:

> Lat take a cat, and fostre him wel with milk,
> And tender flesh, and make his couche of silk;
> And lat him see a mous go by the wal;
> Anon he weyveth milk, and flesh, and al,
> And every deyntee that is in that hous,
> Such appetyt hath he to ete a mous.

But by the thirteenth century, Europeans were hunting and persecuting cats as embodiments of the devil. Ironic as it seems, in the face of what we now know about the Black Death, more attention was devoted to the slaughter of the rat's traditional archenemy than to the destruction of the plague-carrying rat itself. Defoe writes that some 200,000 cats were destroyed in England during the 1665 Great Plague alone because they were thought to carry plague.

Richard Brathwaite wrote a poem about a foolish cat killer in the late seventeenth century:

> To Banbury came I, O profane one!
> Where I saw a Puritane-one
> Hanging of his cat on Monday,
> For killing of a mouse on Sunday.

Frederick the Great made cats official guards of his military food supplies, and by the time of the Industrial Revolution the cat was back in favor because of its value as a ratcatcher in the overcrowded rodent-infested cities of Europe. The same applied the world over, and by the early twentieth century we begin to hear stories of boatloads and planeloads of cats ferried into countries to kill rats, as happened in Bolivia when an outbreak of typhus occurred. Legends about the cat as rat- and mousecatcher grew in every part of the world. In South America, for example, there is the old tale of the cat, mouse, and dog who lived in the same house. The mouse and cat were mortal enemies, of course, but the dog and mouse lived on good terms. So when the mouse heard barking one day he came out of his hole to greet his canine friend. He was quickly caught by the cat. "You're not supposed to bark!" were the rodent's last words. Replied the cat, licking his chops: "To get along in this life, it is necessary to know more than one language."

Among the most highly developed predators in the world, cats have ears sensitive enough to hear the scrabble of rodent paws in their burrows beneath the ground. Just an ordinary cat is fully capable of killing a dozen mice in a short time, often toying with them, playing cat and mouse, before dispatching them. Cats also do strange things with rats, things beyond our comprehension. One domestic cat, observed by a respected zoologist, killed an entire litter of half-grown rats it had found in an overgrown garden and methodically carried each one to a clear space twenty feet away. The cat laid the dead rats in a neat row, where it let them rot, without eating any of them. Even in the wild, this many rats would have been far more than it could have consumed before the rats began to putrify. A clear case of overkill among cats? Or do cats purposely control the rat population as well as feed upon it?

During World War I many cats were sent to France to help rid the trenches of rats, which some doughboys described as the size of Saint Bernards, or at least of fox terriers! One popular parody had an Inspector Kidder from the Isle of Dogs coming over to America to buy 500 trench cats for France. In truth, there actually was an ill-fated plan to send 500 American cats to France, and later, in the midst of the Great Depression, sections of the animal pounds in Paris, Marseille, and Le Havre were set aside to breed special kinds of super cats in an unsuccessful effort to rid France of her rats. Just after World War II the American Feline Society proposed sending one million cats to Europe to help fight the serious problem of rat infestation there. This plan, later abandoned, was opposed by some cat lovers because they believed the hungry Europeans would eat the cats. "There are few cats . . . in Europe today because people have eaten them and have used their skins . . . to make such articles of wear as fur coats, fur hats, muffs, mittens, etc.," wrote one irate cat lover to a newspaper. "Cats are fine and noble animals. They do not deserve such a fate. . . . Civilization would be much further advanced if men could learn to act more like cats than rats."

Though there may not be any rats as big as cats, despite the testimony of some sewer workers, many rats are aggressive enough when cornered to attack the most fearsome cat. "A full-grown brown rat will rip the hide off any cat," says one exterminator, unequivocably. Not long ago the Humane Society of the United States exhorted New York City not to take twelve kittens a Long Islander had contributed to help solve Manhattan's rat problem. "Thousands of stray and starving cats in New York's poorest neighborhoods have already demonstrated themselves ineffectual in the war against rats," the society pleaded. "Please don't add to their numbers." The city returned the kittens. Soon after, it was reported that three cats, supposedly protecting a police station from rodents, fled from a pack of rats.

Cats will sometimes call in other cats when the ratcatching gets too tough for a lone cat to handle. In one case, a large tomcat was employed in a 1957 ratting party in an outhouse. Men ripped up the floorboards and fetched the cat, but it darted out the door when it saw the great number of rats below. The cat was dismissed as a coward until it returned in a moment with two younger cats, the three-cat team killing the entire pack.

The most vaunted rodent killer on record appears to be a tabby named Micky owned by the English firm of Shepherd & Sons Ltd. in Burscough, Lancashire, which killed some twenty-two thousand mice between 1905 and 1928. But the truth is that cats—which, incidentally, seem to bite more people than do rats every year (871 cat bites and 255 rat bites were reported in New York City during 1979)—are not especially valuable in reducing the number of rats in the world. The most reliable study done to date found that in residential areas the presence or absence of cats had no measurable effect on the number of rats. The conclusion was that cats "ate only about 20 percent of the rats that must die anyway to maintain a stationary population."

Another study, however, found that farm cats ate enough rats to prevent the expected upsurge in the spring rat population, it being apparent that those cats ate rats even when cat food was present: scientific proof of Chaucer's poem! But by and large, research indicates that the only substantial effect on a rat population by cats is to limit its spread to new areas. Humankind still has great faith in cats, though: of the 4.9 million cats in Great Britain some 100,000 are employed by the civil service as ratters and mousers. Well-trained cats probably do help in the war on rats.

Our expression "to ferret out," to search or dig out relentlessly, comes to us from a relative of the weasel called the ferret, which farmers still use to rid their fields of destructive rats and which was once used in cities throughout the world. A natural enemy of rats, this nocturnal animal is trained and kept in cages. When a rathole is found, the pocket-sized ferret is placed before it. The slender slick-coated creature, "fury in fur," crawls through the underground passages burrowed by the rat, its pink eyes able to see in almost complete darkness. Snarling savagely on sighting a rat, the ferret swipes his paw at it and the rat strikes back, but it soon realizes it has no chance against the ferocious predator. The rat flees toward daylight, the ferret close behind him, and as soon as it emerges from the hole the ferret leaps up and rips open its throat. Not many ferrets are used in cities today, but as recently as 1928 there were several "ferreters" in downtown New York, including one on Fulton Street with a window display of stuffed ferrets in battle.

The mongoose, like the ferret, is a natural predator of the rat, but though it needs no expensive training or care it causes far more trouble than the ferret does, more trouble than it is worth. "Please never introduce the mongoose into this country," G. A. Seamon pleaded after presenting a paper on the animal to the North American Wildlife Conference in 1952. Reviewing mongoose history, one can understand his plea. West Indian planters thought they had solved the black rat problem in 1872 when they imported nine mongooses from their native

Nepal. Black rats were destroying a quarter of the sugarcane crops annually by nibbling the stalks, which lets in fungi, causing the sugar to ferment and ruining the entire plant. At first this figure was drastically reduced by mongooses, the reduction in rat damage on Jamaica alone exceeding ten thousand pounds annually within a decade. But it quickly became clear that rats were not enough for the greedy, catlike mongoose, which began to prey upon birds and other animals. To date, the mongoose has exterminated or virtually exterminated the Jamaican nighthawk, the burrowing owl, the black-capped petrel, the tree duck, the native iguana, eight forms of native lizards, and has even been raiding nests of green turtles and pelicans for eggs. It also preys upon baby deer, snatching the fawns by the nose.

Meanwhile, the rats have become as numerous as ever on most West Indian islands, still ruining 25 percent of the sugar crop. Today, six islands offer bounties for mongoose bodies! It is much the same in parts of northern South America and on several Pacific islands, and even worse in the Hawaiian Islands. Thirty-six pairs of mongooses were introduced to Hawaii in 1883 and others two years later, the animals substantially reducing rat damage to the sugarcane crop. Tests of the contents of mongoose feces showed that 52 percent contained nothing but rat and mice remains, while the rest was composed of rodent and insect remains. It developed, however, that the mongoose needs more food than a rat diet can provide, primarily because the rat is a night feeder and the mongoose feeds by day, their paths crossing too seldom to feed a large mongoose population. While it remained the only large-scale predator of the rat in Hawaii, the mongoose exterminated or seriously depleted populations of petrels, Newell's shearwater, red-footed boobies, native geese, doves and lizards. The Hawaiian black rats are still flourishing, thanks to their night-feeding habit, their high reproduction rate and the mongoose's inability to catch this arboreal animal when it takes to the cocoa trees and uses their interlocking branches as runways. So much for the mongoose, especially when we consider that mongooses carry rabies, can be infected with Weil's disease or infectious jaundice like the rat (which probably infects them with it), are often infested with the rat fleas that cause bubonic plague, and can harbor the roundworm *Trichinella spiralis,* which causes trichinosis in people who eat the wild pigs that eat infected mongoose flesh.

Even ants have been employed by people trying to eliminate the mighty rat; Thomas Raffles introduced Cuban ants to the West Indies in 1762 in the futile hope that they would destroy enough young rats in the nest to control the rat problem. In early frontier America, minks were sold as ratters. Skunks, weasels, leopards, foxes, lizards, giant toads, various mammals and birds of prey such as hawks all hunt and eat rats. "Rat snakes" in Ceylon are often so domesticated that they are fed at the table. Snakes like the Indian python crush and gulp down rats and mice, though there is at least one recorded case of a scrappy rodent who bit a South Australian inland taipan while being swallowed and caused an infected internal wound that almost killed the snake. (According to snake owners, rats fed as food to their pets often bite the snakes.)

The little narrow-footed marsupial mouse, hardly bigger than a true mouse, is also a rat killer. So is the great horned owl, which sometimes feeds heavily on rats in Iowa cornfields, although owls usually feed on wild rodents and have several times decimated bird populations where introduced. In 1927, the United States shipped a large number of native owls to the South Sea Islands to exterminate rats where cats had conspicuously failed. A man in Morristown, New Jersey, is presently trying to raise owls in a local church belfry to use as rat and mouse killers, while the small city of Satellite Beach, Florida, uses both red-tailed hawks and owls to keep down its rat population: a dozen or so hawks work the day shift, catching rats, and a similar number of owls work the night shift. Owls and many other species of rat-catching birds, however, often die from eating dead rats poisoned by man.

Some years back, keepers at New York's Central Park Zoo found that alligators were the best rattraps they had. Within a few days, it was reported, alligators had been observed killing about a dozen rats. An observer told how it invariably happened:

> The big alligator was lying near the edge of the pond with his jaws wide open. His eyes were shut tight, and I wondered if he had gone to sleep with his mouth open. The other alligators were all asleep on the other side of the cage. Some of them were in the water. There were two rats nosing around in the pen, feeding on some pieces of meat that had been left by the alligators. There must have been a piece of meat in the big alligator's mouth, although I did not see it there, for while I was watching one of the rats went over and began smelling around the big alligator's jaws. He put one foot on the alligator's lip, hesitated an instant, then reached his head way over, as though he were after something. Suddenly the alligator's jaws closed with a snap, and his teeth caught the rat just back of the neck. He gave some kicks and then the alligator began to chew him up.

It is said that pet alligators flushed down toilets by their owners when they tire of them, grow large in the sewers and prey upon rats, helping keep their numbers down. But conservative estimates indicate there are still 500 rats to every mile of sewer.

In the San Francisco bubonic plague epidemic of 1906, Danysz's Virus, which produces a mild typhoid fever fatal to rats and harmless to men, was used to kill rats, and a bacterium called azoa was said to be used to cause a deadly disease among the rodents, too. The role of pathogens in limiting rat populations has been extensively studied. Scientists now know that some disease organisms, such

This Elizabethan woodcut shows a ratcatcher advertising his services.

An English ratcatcher drawn by William Collins in the nineteenth century. *(N.Y. Public Library Picture Collection)*

(above) A ferret finishes off a rat. *(Richard Lydekker, The New Natural History)*

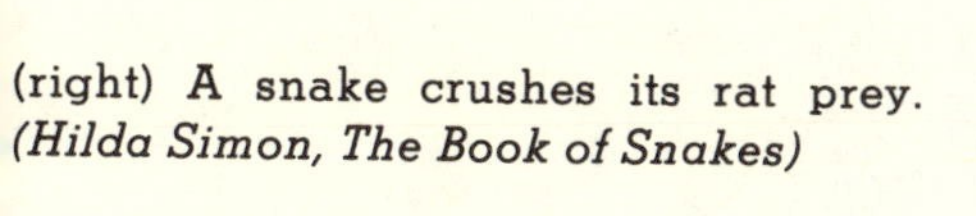

(right) A snake crushes its rat prey. *(Hilda Simon, The Book of Snakes)*

(left) A gnome owl with its prey *(Richard Lydekker, The New Natural History)*

(above) Crested cassiques, another of the rat's many feathered enemies. *(Richard Lydekker, The New Natural History)*

(right) Egyptian mongooses have killed many rats but cause even worse problems themselves. *(J. G. Wood, Animated Creations)*

A bullterrier, a fierce and courageous breed, often used in ratting matches *(Dogs of Today)*

(below) At ratting matches, popular in America and Europe until they were outlawed in the early 1900s, dogs were pitted against rats to see which dog could kill the most rats in the shortest time. *(Mayhew's Life and Labour of the London Poor, 1849, N.Y. Public Library Picture Collection)*

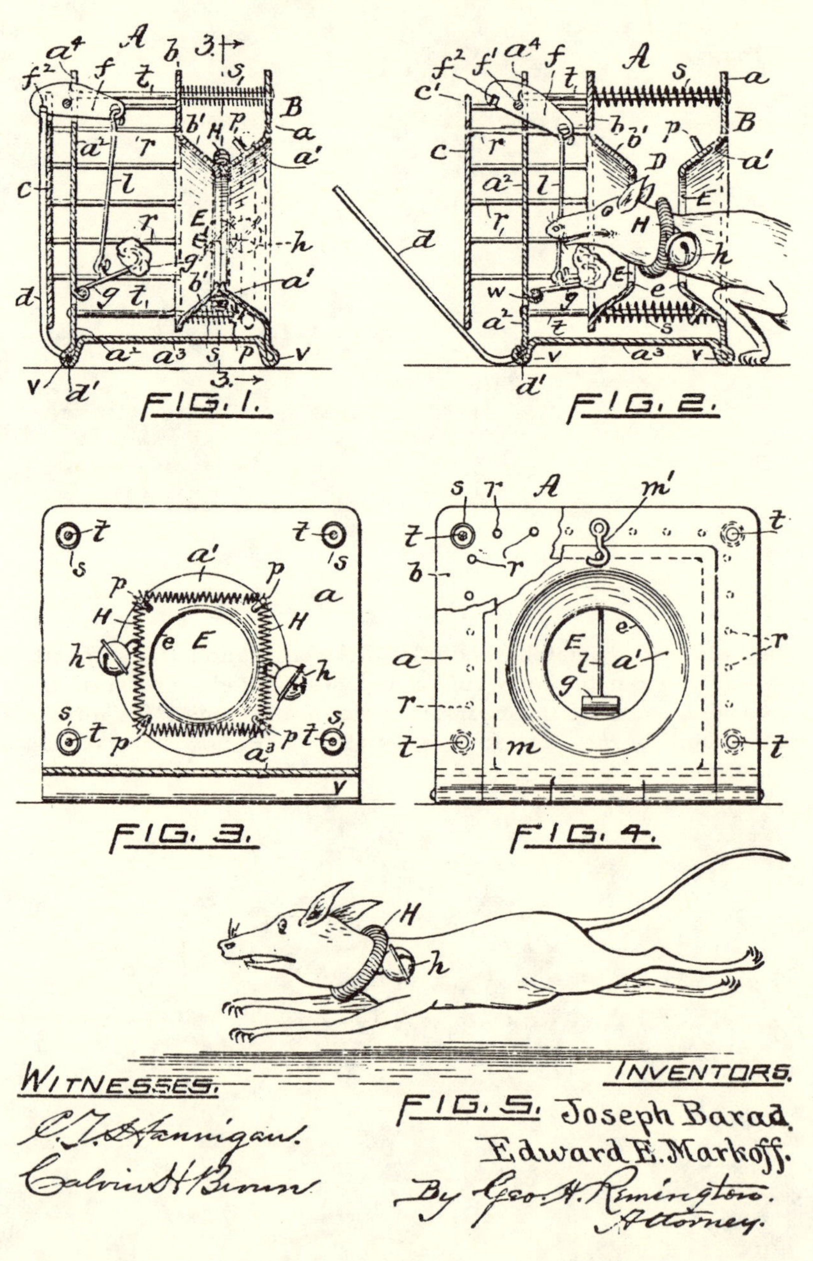


This illustration is part of the patent, granted in 1908, for the "Humane Mousetrap." In short, the device was designed to clamp a bell collar around the rodent's neck and set the animal free, thereby frightening away all the other mice when the now noisy mouse returned to his pack. *(U.S. Patent Office)*

as the bacterium *Yersinia pestis*, which causes plague, are highly fatal to rats, while others, such as *Rickettsia typhi*, the agent of murine typhus fever, are rarely fatal to the rodent. The spirochete which causes Weil's disease in humans has little or no effect on rat populations, and while *Salmonella* bacteria appear to cause high winter mortality of rats, the many human attempts to use it to control the rodents have had little long-range value.

Man has always been the best rat killer among the animals. Except for the Pied Piper, ancient ratcatchers were on the whole far less successful than their modern counterparts (see chapter 6), but then they were somewhat less scientific. An ancient Greek farming manual gives the following formula for ridding the premises of rats, should someone be interested in trying it. Simply take a blank piece of paper and write down: "I adjure you, ye rats or mice here present that ye neither injure me nor suffer another rat or mouse to do so. I give you yonder field [here a particular field is indicated], but if I ever catch you out of it and here again, by the mother of the Gods I will rend you in seven pieces." The paper was to be placed on an unhewn stone, "taking great care to keep the written side up" so the rodents could read it without going to the trouble of turning it over.

Centuries later the Welsh used a similar method, without the trade-off, but their magic words had to be "put into the mouth of King Rat," no instructions given as to where or how to find him. The magic inscription, should anyone want to look for King Rat, reads as follows:

r.a.t.s.
a.r.s.t.
t.s.r.a.
s.t.a.r.

The ancient Greeks placed great faith in their god Apollo Smintheus as a ratcatcher, calling him "Ratkiller." He earned this sobriquet when he sent a swarm of rats against his priest Crinis because Crinis had neglected his duties. When the priest repented, seeing the invaders coming, Apollo forgave him and with his far-reaching arrows killed the swarms of rats he had sent.

The only Biblical mention of rodent prevention concerns the plague-stricken Philistines, whose priests and diviners called for the return of the Ark of the Covenant to Canaan with the "images of your mice that mar the land." (*I Samuel* 6:5). Similarly, the Indian *Bhagavata Parana* made no specific mention of ratcatchers, but warned that it was time for men to leave their homes when a rat fell from the roof, reeled drunkenly on the floor and died, "for then be sure that the plague is at hand."

That the early Chinese were aware of the health dangers of rats is witnessed by the admonition of a first-century physician who warned that "if rats have run over the rice basket, throw the rice away and don't eat it." Bishops of the early Roman Catholic Church regularly prayed for St. Gertrude's protection against rodents and plague. St. Gertrude (d.664), who is also the patron saint of travelers, still remains the saint to whom one should pray. The former Abbess is sometimes represented in religious paintings as surrounded by rats or with rats and mice running about her distaff as she spins.

Jews in fifteenth-century Germany often served as ratcatchers in return for exemption from some of the many restrictions to which the great majority of Jews were subjected. There were so many ratcatchers in Europe during the Middle Ages that they were organized into guilds and by Shakespeare's time a proficient ratcatcher might be an honored citizen of his town. One perquisite of the job were the rat pelts a ratcatcher could accumulate, for soft black rat's fur was much used in the Middle Ages in making inexpensive coats or as trim for garments. Later, during an English rat-killing crusade early in the nineteenth century, the rat skins obtained were valued at almost a quarter of a million dollars a year, being used in bookbinding, picture frames, purses and even for the thumbs of ladies' gloves. "Rat leather" was similarly used in Japan a century later and is still used in France to make ladies' gloves.

Medieval ratcatchers wore uniforms or carried standards designed to attract attention as advertisements for their services. They were organized into guilds with rat poison peddlers, some of them rich men. Chinese ratcatchers carried a cat in a bag as the ensign of their craft. The seventeenth-century Italian ratcatcher, as depicted in Annibale Caracci's illustrations of the *Cries of Bologna,* shouldered a pole bearing a square flag on which there were pictures of rats and mice. The most famous representation of one of these medieval Pied Pipers is by the seventeenth-century Dutch engraver Cornelius Vischer, who depicted a peg-legged Dutch ratcatcher in a colorful costume carrying a rat flag and with dead rats hanging from his high hat, cuffs, sword handle and belt. Vischer even inscribed this ratcatcher's cry on the humorous print:

> By the cat you put rats to flight. If
> you drive away little thieves by great
> ones, it is utter folly. Look at me;
> provided only a little coin is forthcoming,
> I will put both rats and cats to flight.

There is some evidence that the tale of the medieval Pied Piper of Hamelin, who led the town's children away when cheated, is based on a very real human ratcatcher (see chapter 7). In any case, ratcatchers were common on the streets of Europe. In Elizabethan London their cry was a gay one full of the energy and poetry of Elizabethan language:

Rats or mice! Ha' ye any rats,
mice, polecats or weasels.
Or ha' ye any old sows
sick of the measles?
I can kill them, and I can kill moles,
And I can kill vermin that creepeth up
and down and peepeth into holes!

The Irish, who could also turn a phrase, believed that they could rhyme a rat to death if the poem made was good enough. It was commonly believed in Ireland that rats could be dispatched by anathematizing or cursing them in rhyming verse or by metrical charms. Ben Jonson alludes to this when in the *Poetaster* he writes "Rhyme them to death as they do Irish rats," and Shakespeare alludes to the belief in *As You Like It,* making Rosalind say: "I was never so be-rhymed since . . . I was an Irish rat," in referring to the Pythagorean doctrine of the transmigration of souls.

Throughout Europe it was once thought that rats would evacuate a house through the remaining corner if the Gospel of St. John was read from the other three corners. Pieces of cloth sprinkled with holy water placed in three corners of the house supposedly worked just as well.

It was also widely believed that red-haired men were so deceitful that the fat of their bodies made one of the best rat poisons. This was almost as absurdly good as the cannon of Sweden's Queen Christina (1626–89). Christina so hated the fleas of rats and other animals that she officially declared war on them. She packed tiny cannonballs into a specially-made four-inch cannon, today on exhibit in the Stockholm Arsenal, and fired at the fleas whenever they came by. Others shared her dislike of fleas. One F. E. Bruckman wrote a ninety-four-page book on them in 1739 in which he told how to catch fleas and dispatch them "by murder, beheading, hanging, or some similar end."

From the Middle Ages until about the late eighteenth century it was thought that rats and other pestiferous animals could be exterminated by trying them in courts of law. Many such cases are recorded in E. R. Evans' *The Criminal Prosecution and Capital Punishment of Animals* (1906). Wild animals such as rats fell under the jurisdiction of the ecclesiastical courts under the reasoning that "as God cursed the serpent, David the mountains of Gilboa, and our Savior the barren fig tree; so, in like manner, the church had full power and authority to exorcise, and anathematise, and excommunicate all animals and inanimate things." When rats inundated a district, they were summoned to court and when they failed to appear, they were ordered to leave the district, depart to a foreign country or swim out to sea. If the rats didn't obey and became worse than ever in the area instead of "withdrawing off the face of the earth" as commanded, this was attributed to the malevolent antagonism of Satan, "who is at certain times permitted to tempt and annoy mankind."

Summonses in such cases were served by an officer of the court who read them at the places where the rats were last seen. Written in all technical formality, they also contained a full description of the rats, so that no mistake might be made. In a process against rats in Autun, France, for example, the defendants were described as "dirty animals in the form of rats, of a grayish color, living in holes." It was in this particular case that the celebrated French lawyer Chassanée, known as the Coke of France, won his first laurels. Chassanée was appointed as an advocate to defend the rats, as was the usual procedure. When the rats didn't appear on the first citation, he argued that the summons was of a too local and individual character, that, as all the rats in the diocese were involved, all the rats should be summonsed in all parts of the diocese. This plea was accepted and the curate of every parish was instructed to summon every rat for a future day. When that day came, bringing no rats, Chassanée claimed that his clients were summoned too hastily—they needed more time to prepare. Another extension was granted but still no rats appeared. Chassanée then changed his strategy, as a biographer later noted:

> Chassanée objected to the legality of the summons.... A summons from the court, he argued, implied full protection to the parties summoned, both on their way to it and on their return home; but his clients, the rats, though most anxious to appear in obedience to the court, did not dare to stir out of their holes on account of the number of evil-disposed cats kept by the plaintiffs. Let the latter, he continued, enter into bonds, under heavy pecuniary penalties, that their cats shall not molest my clients, and the summons will be at once obeyed. The court acknowledged the validity of this plea; but the plaintiffs declining to be bound over for the good behavior of their cats, the period for the rats' attendance was adjourned *sine die,* and thus, Chassanée gaining his cause, laid the foundation of his future fame.

Superstitious behavior prevailed among ratcatchers until well into the nineteenth century, when one recipe for destroying rats called for catching a rat, brushing it with tar and oil, and letting it escape into the walls, where its horrible fate would scare off all the others. Similarly, people believed that setting one rat afire aboard a ship would cause all the other rats present to jump overboard, and that a captured rat released after having its anus sewn up would frighten away the rest of the pack. All of these cruel methods had their basis in the belief that the screams of a rat in pain would drive off its companions. More humane, but equally ineffective, were those householders in Germany who paraded through their homes banging pots and pans in the hope that the noise would frighten the rats away.

Another, more practical, old ratcatching recipe instructed the householder to cut corks thin as sixpence, stew them in grease, and place them in rat tracks so that they would stick to a rat's fur and "cause its departure." In place of the cork in grease one could use dried sponge dipped in honey, oil of rhodium, or bird lime. But as early as 1768 scientific methods were being employed in the war against rats. Robert Smith, a prominent ratcatcher, published a book that year in which he instructed "the gentleman, farmer, and warrener" to leave traps unset at first "in order to embolden the rats before you set the traps in earnest to catch them." Even today it is recommended that rats be given nonpoisonous bait to win their confidence before poisonous baits are used.

The nineteenth century saw the development of the first reasonably effective modern rat poisons as we'll see later. Some celebrated children of the century engaged in killing rats. Darwin, for instance, participated in ratcatching, and Marcel Proust, according to George Painter's recent biography, showed his hatred of them by his sadistic practice of hiring "young men to put on exhibitions in which they chased rats about a room to stick them with needles and beat them with clubs." Apparently his sensitivity to the slightest cruelty shown to living things, as amply recorded in his novels, did not extend to rats. Most rat haters have not been so sadistic. Alfred I. du Pont of the millionaire Delaware clan was content with inventing an electric rattrap. The champion of the era, one Emile Dusaussois, who operated a slaughterhouse for horses near Paris in the 1800s, simply used horsemeat bait in ordinary traps to kill more than 1600 rats in one month.

During World War I, rats were a terrible problem in the trenches, as Erich Maria Remarque makes quite clear in *All Quiet on the Western Front.* Remarque tells of rats that bit two large cats and a dog to death, rats growing fat eating human corpses in No Man's Land, and swarms of fleeing rats storming the trenches. The men, of necessity, became expert rat killers. In one instance they heaped scraps of bread in the middle of the floor and lay down flat on the ground. When they heard "the sound of many little feet," they switched on their flashlights, stunning the rats, and beat them to death with their spades.

A rat control measure much used in both ancient and modern times is the bounty for dead rats. A Jew in fifteenth-century Frankfurt, for example, received special privileges if he turned in five thousand rat tails a year. Recently, a Chicago alderman offered his constituents a $1 bounty for each rat captured. Often fantastic numbers of rats have been killed as the result of bounties of cash, or honor in one's land or ideology. Over 12 million rats were exterminated in India in 1881 for rewards of one kind or another, but the record is probably the 1.5 *billion* rats caught by Chinese civilians after Mao Tse-tung launched an antirat campaign in 1952—about two rats per person.

Rat trappers during the 1906 San Francisco bubonic plague outbreak weren't catching enough rats until the city paid them a bonus for rats caught, in addition to a straight salary. Chewing gum, rather than cash, inspired children on an unidentified South Pacific island during World War II; the native kids killed fifteen thousand rats, for which the U.S. naval chaplain gave them thirty thousand sticks of gum, or two sticks for each dead rat. At about the same time in Geneva, New York, a thirteen-year-old girl won a twenty-five dollar savings bond for killing fifty-four rats, which may be the American kid's record but doesn't compare to the efforts of her South Pacific island counterparts.

But bounties can lead to problems. In 1947, a keeper in the Bronx Zoo who got a twenty-five cent bounty a head for shooting rats as part of the zoo's rat-shooting squad (used because poisoning might endanger valuable animals) shot and wounded another keeper while zealously pursuing his duties.

Human ratcatchers have come up with a great number of novel methods for killing rats. Back in the 'twenties the family flivver was recommended as a rat exterminator. Department of Agriculture experts suggested hooking up a hose to the exhaust pipe, inserting it into a rat burrow, starting the motor, and shooting the carbon monoxide into the burrows, where it would be effective up to sixty feet away. This is more ingenious than the scheme an Italian seaman aboard the passenger liner *Dante Alighieri* had at almost the same time. He lured rats into the storeroom by playing enticing Italian melodies on his piccolo and had his partner blast them to bits with a pistol, at considerable damage to the food, which included a huge Italian cheese that finally resembled the Swiss variety. Yet neither scheme could rival the following 1878 tongue-in-cheek invention of a "cornucopia" for killing rats described in a New York newspaper:

> They are sold in two different shapes, either of which would be useless without the other. The rat cornucopia No. 1 is simply a hollow cone formed of white paper, and a trifle smaller at the base than is the average rat's head. A few particles of cheese are placed in the smaller end of the cone, and it is then laid on its side in a closet especially popular among rats. Of course, the cheese vanishes in a very short time, and in the course of two or three days the rats become so thoroughly convinced that the cheese-bearing cornucopia is an honestly meant tribute to their merits that they discard every suspicion that it bears any relation to traps. Now is the time to use cornucopia No. 2, which resembles No. 1 with the important exception that its interior is lined with deodorized coal-tar. When a hungry and trustful rat approaches this apparently innocent device and squeezes his head into it in order to seize the cheese, he finds that he cannot withdraw. The inexorable tar has glued his head fast to the encircling cornucopia and he is virtually blind. When a rat has thus "bonneted" himself, his first impulse is to fly to his apartments in the partition or under the floor; but he is utterly unable to find the entrance hole. Terrified by his situation he rushes recklessly about the shelf, upsetting all sorts of things, and creating an uproar that instantly summons the head of the house,

with a large pair of tongs, with which weapon the rat is easily captured. At this point, short-sighted people would kill the rat, but science shows a better way. By the aid of the tongs, the head of the rat is inserted into the nearest rat-hole, and he returns to the rat community in the character of a dangerous and unusual ghost. His family renounces him, and his nearest rat friends at sight of his white, featureless and conical head, utter squeaks of horror and rush madly away. Within an hour after this catastrophe, every other rat will have fled from the haunted premises and sought shelter at the next house. The unhappy victim of the cornucopia, writhing under the false accusation that he is a ghost, and unable to eat or drink, perishes miserably. Thus the house is freed from rats, and the householder calls his wife together and bids her rejoice with him over the defeat of the enemy.

But the most complicated, impractical and useless of all ratcatching inventions must be the Humane Mousetrap given U.S. Patent number 883,611 by the U.S. Department of Commerce Patent and Trademark Office, which describes the device this way: "When a mouse sticks its head into the Humane Mousetrap for a piece of cheese, the trap snaps shut, clamps a bell collar around the rodent's neck, and sets it free, unhurt. The trap becomes effective when the animal returns to its home colony and announces its coming by the sounds emitted by the bell, thereby frightening the other rats and causing them to flee, thus practically exterminating them in a sure and economical manner." So far nobody is manufacturing the Humane Mousetrap.

It was hoped at one point that rats themselves might really turn out to be the greatest ratcatchers or rat killers. In 1967, rat mutations were discovered producing sterile rats that continued to copulate constantly, as is the wont of male rats. Experts suggested that such rats be bred and released like a "fifth column" into the population of wild Norway rat populations. Scientists also held that laboratory-bred sterile rats would be even more effective if they were "feralized" or made wild before being released into a strange environment, that highly aggressive sterile male rats should be developed to attack and kill wild males on release, and that the sterile males released should also have poison resistance bred into them so that they would live longer than the wild rats. As with many plans of mice and men, however, the trouble was that none of these theories worked. In 1979, Dr. Allen Stanley, who "invented" the so-called Super Sterile Rat, confirmed that every such sterile rat placed into a wild colony was promptly killed after being attacked by both male and female resident rats, who then consumed their carcasses. The intruders, like all rat strangers, offered no resistance to the attacks, often exposing the neck and vital jugular vein to attackers. It was also pointed out by skeptics that since every female rat in a colony entering the estrous stage of her reproductive cycle mates with every mature rat in the colony, a colony would keep growing where even one fertile male rat remained.

In any event, the Super Sterile Rat turned out to be not much more than rat food.

It is astounding, almost as revolting as the rat can be, that it took nearly half a century from the time when experts knew how to combat rats until a concentrated effort was made throughout the world to wage modern scientific war on the destructive rodents. For years, the language of evasion and the politics of postponement prevailed. Little more was done than in those poor countries where the peasants buy special masses to pray for protection from rats, or embark on rat hunting forays clubbing rats to death. Indeed, back in 1909 an Indiana legislator proposed an annual "Rat-killing Day" when the state would compel residents to help exterminate the pests or pay a stiff penalty for failing to do so. In Berlin, Germany, on November 20, 1926, authorities actually declared a three-day war on rats, instructing citizens to use poison, popguns, small rifles, fox terriers, rattraps, poisons, and any other feasible methods for killing the pernicious vermin. The call to arms was not merely an appeal but an emphatic injunction to all house and landowners to kill rats. Disregard of the order was punishable by a fine of up to 100 marks whether rats were observed on the premises or not. For three days rats never had it so thoroughly and methodically bad in Berlin.

In Java, several years later, authorities declared that anyone wanting to marry had to pay twenty-five rats' tails for a license; when as a result an artificial rats' tail industry developed, the authorities changed the requirement to twenty-five dead rats, whereupon a number of rat breeding farms were founded. Other official rat-killing schemes called for exchanging ten dead rats for a movie ticket, fining landlords for not exterminating rats, and even fining lax city agencies. "I am amazed and astounded by the rat-bite cases that are told me from day to day," a New Jersey judge said in 1924. "It has reached the stage where mothers are afraid to leave their babies long enough to go into the street. Complaints are made to the Health and Tenement House Departments. They report the violation to the landlords and that is the end of it. They do not enforce the regulations. If they did, they could control the plague of rats."

Most of the rat remedies were "like giving an aspirin for cancer," as one exterminator put it. Some of the fault lay with people who were so apathetic, disillusioned, ignorant, or criminally irresponsible that they "took rat bites of their own children as a matter of course, describing the incident as if it were a joke," in the words of an investigator, "unless the child is so badly hurt they must carry it to the hospital." People seemed unaware of their rights as human beings and government unaware or uncaring of its responsibility to the people.

It was finally rats biting children, more than any other single factor, that led to the passage of the national legislation in the United States that at last made possible the modern, scientific war against rats described in the following chapter. In the 1960s black activists, including Jesse Gray, the Harlem rent-strike leader, and Dr. Martin Luther King, Jr., led the fight for effective national direction in the war against rats, which in turn would inspire action in other nations around the world. Gray and eight other demonstrators were arrested in the House of Representatives gallery during the summer of 1967 as they chanted "Rats cause riots! We want a rat bill!" A bitter political struggle commenced, during which Congress voted down President Johnson's first bill for a $40 billion two-year effort, and Johnson observed that Congress was doing more to protect cattle than American children.

The "Civil Rats" battle raged for some six months before any action was taken. While their colleagues snickered, Neanderthal Congressmen jested about "rat patronage" and "rat bureaucracies," deploring the creation of a "rat corps" headed by a "high commissioner of rats." But eventually, on December 5, 1967, Congress passed a national health insurance bill that included the first federal spending for rat control aid to the cities, although the states were authorized to decide whether the money actually should be spent for that purpose. Clearly, media coverage of rat-bite stories, advertising about the awful rat-bite incidents in the ghettos, and demonstrations such as the one in which an elected official was encountered by activists dangling freshly killed rats by their tails, had made the legislation possible. The psychological impact of rat bite, the fear of parents lying awake at night listening to scraping sounds in walls and fearing their children might be bitten, turned the tide in the battle. As New York's Governor Nelson Rockefeller said, "The tragedy of what's been going on—the child who's having his toes or ears chewed off at night—became a symbol of the lack of concern for human value!"

# VI

# THE WAR AGAINST RATS

The woman in black had worked late that night and walked slowly to her car parked just south of City Hall. It was a warm mid-May evening and she seemed to be enjoying her leisurely twilight stroll after a day cramped behind the desk. Then she felt a curious tug at her leg.

The tug became a sharp searing pain and she cried out, startled, but when she looked down she was almost shocked into silence.

A huge rat was biting her leg, and just behind it a swarm of rats—at least thirty of them—were nearly upon her.

She kicked the matted brown thing off her, kept kicking and screeching as the rats darted in and out seeking vulnerable flesh, several wrapping themselves around her legs and picking at her until her screams alerted a man down the street who came running to her aid.

The young man pulled off his light summer jacket and tried beating off the rat pack, but his efforts were useless—more rats were advancing out of dark burrows in the vacant lot bordering the street. He saw that the woman, in a "state of hysteria," did manage to break free into her car and he ran from the rat swarm to call for emergency police aid.

The rats kept coming. Scores of rats blanketed the sidewalk when the police arrived a few minutes later, and they did not cease their strange aggressive behavior until patrolmen opened fire, killing several of them. The remainder fled into the lot.

Eyewitnesses said the mysterious woman in black sped off in her car still screaming, but she was never seen or heard from again after that night of May 10, 1979, even though New York City officials pleaded for her to come forth, fearing that she might be carrying a serious ratborne disease and be in need of prompt medical attention.

More mysterious was the aggressive behavior of the rats. Did the pack merely panic and "stampede" when the woman inadvertently walked into their midst, or did they attack without provocation, a strange breed with a new, dangerous behavior pattern? Possibly the former is true. "It's very unusual for rats to make unprovoked attacks on people," says New York City Deputy Health Commissioner Jean Crapper. "They're not interested in confrontations. But if they did attack this woman unprovoked that would be very serious and cause us to rethink many of our present rat-control procedures."

Whatever the case, the New York City Bureau of Pest Control was taking no chances. The next morning, teams of Bureau exterminators in orange helmets and blue uniforms were at the attack site, the vacant L-shaped lot once the home of Ryan's Cafe, which had been razed nine years before after a gas explosion in which twelve persons were killed. The block was roped off and the rat kill began. As a first step, the "foot soldiers of the rat wars" removed from the lot the piles of garbage that, for almost a decade, had been illegally dumped over the fence by private carters and nearby fast-food restaurants. Without food to eat, the rats would be more likely to sample the quick-acting zinc phosphate poison baits injected into their burrows. More than two tons of garbage were removed, scores of burrows and rattraps were baited, and a giant rattrap built of wire mesh, 15 feet high and 25 feet wide, was filled with the "Rat's Delight"—peanut-butter sandwiches topped with poison.

By Monday, over 100 of the estimated 300 rats in the lot had succumbed to the poison-spiked peanut butter in the giant rattrap. Another 50 to 100 had died in their burrows, many of them lying dead against the fence, their eyes now red and open, bellies pink in the sun, scaly tails raw and dripping. It was time for the exterminators to wade in and clean up the surviving rats.

The orange-helmeted teams began digging into the remaining pile of decayed garbage, rotted wood and rusted metal beer cans. "Over there!" a worker shouted, and another man smashed a darting rat with his shovel, stunned it, scooped it up and smashed it against a wall. A second rat exploded up and a worker deftly impaled it on the tines of a rake. Exterminators were killing rats wherever they poked and dug in the lot. One worker killed five in a row with a large spiked board, rats boiling up out of the burrows. "They're babies—we hit a nest of them!" someone shouted, and the rat killers worked at a frenetic, deadly pace—*whack, whack,* a pathetic squeal, *pthump, splat,* rat flesh and fur flying in the air until the nest was wiped out, the exterminators lifting each dead rat by the tail with gloved hands and casting it into a pile that grew four feet high before they were finished.

There was more to do: more poison had to be spread in the general area, and ratholes ("It's a poor rat that has only one hole to run to," as the proverb goes) had to be filled with a cement-glass mixture so that any surviving rats would kill themselves trying to gnaw their way to freedom. But Randy Dupree, director of the Pest Control Bureau, was satisfied that his foot soldiers had cleaned this lot of rats. One of the rat killers, face glistening with sweat, wasn't so sure. He had grown up in a rat-infested apartment, he said, and he got satisfaction killing rats, though little satisfaction in the $6,599 to $7,200 a year a city exterminator makes. "Downtown here, some of these buildings have double basements," he said. "There's tunnels from the Revolutionary War, which I've seen on old maps, tunnels that were sealed up 200 years ago. There's rats in them. There's underground streams, and rats in them, too. Hundreds of thousands of them. They just breed. And if they got nothing to eat, they eat each other. . . . They'll eat their young, they'll eat anything. And they're down there. Millions of them. All down there."

He speaks without much passion, almost matter-of-factly. A reporter tries to get a rise out of him.

"What do you dream about at night?" he asks.

"Girls," the foot soldier says, laughing, and he goes back to killing rats.

They are the same the world over, the rat killers, hardworking, even courageous men, not a bit squeamish about their work or given to much introspection about killing rats. And they do a good job. In New York City, for example, where rats are very democratic, settling in every kind of neighborhood ("I have seen rats that like spaghetti and meatballs, rats that like corned beef and cabbage, rats that like soul food, rats that like *arroz con pollo,* and rats that prefer bagels and lox," says a local exterminator), the war on rats has been markedly successful, as it has been in Chicago, Miami, and many other U.S. cities.

In the Rockaways section of New York, this rat reporter watched another successful operation, in which garbage-strewn vacant lots were cleaned, and waxy green poison, "Rat Bait Blocks with Fish Flavoring," was positioned under the five-mile-long boardwalk, after the end of beach season in that resort community. Prominent "Rat Poison" warning signs were posted at the sites as always. New York exterminators also cover the waterfront docks; and health authorities carefully inspect ships in the great port city, looking for any rats that might be trying to immigrate, and seeing that precautions (such as dish-shaped guards on mooring lines) are taken against their landing.

Success in controlling seafaring rats and their stowaway fleas is one of the primary reasons for the decline in ratborne epidemics such as bubonic plague.

Modern shipbuilding, the education of shipowners and seamen, and strict inspections of all ships entering ports are responsible for this major victory in the war against rats. Rat guards alone are hardly enough: rats can easily leap over the 36-inch-diameter mooring-line guards, walk for sixty feet over a quarter-inch cable, or swim away from a ship and scramble up a perpendicular bulkhead, clinging only to the thickness of the paint. Sanitary inspectors, in countries throughout the world, must first thoroughly inspect a ship at a quarantine station *before it pulls into port* and fumigate it if rats are found aboard. If rats, or signs of them are discovered tear gas bombs are thrown into the vessel to flush out any human stowaways before a more lethal gas is used to kill the rats. This is most important because, in the days before such precautions were taken, stowaways were often killed along with the rats upon release of the deadly hydrocyanic gas. "A few weeks ago, in the hold of a South American freighter, the tear gas brought out eight weeping stowaways who had been hiding in an empty water tank," a New York port official admitted during World War II. "Two fellows in the crew had smuggled them aboard in Buenos Aires and had been feeding them. These fellows had kept their mouths shut and gone ashore, leaving the stowaways to be killed for all they cared."

Even lifeboats are fumigated when a ship is gassed, for rats have been known to hide everywhere aboard a vessel. Such diligent practices have kept rat immigration to a minimum in recent years. No longer do as many as six hundred rats, the New York Harbor record, come in aboard a ship, and very few make it into the cities.

But every city, or "Ratropolis," as some knowing rodentologists would have it, has its rat problems, often in the most unlikely places. Several years ago at one of New York's poshest points, Park Avenue between 58th and 59th, a colony of about 100 rats constructed a labyrinth of burrows underground, and astounded pedestrians by regularly assembling every morning to feast on food left for the pigeons. New York rats still burrow by the thousands in Central Park, where there is plentiful waste. In high-crime areas like Harlem, there remains a serious problem: residents "airmail" their garbage (throw it out the window) because they are afraid to use the basement incinerator room, where a mugger may be lurking, and thus make actual garbage dumps of alleyways. Nevertheless, great progress is being made in the war on rats here and throughout the United States. For one thing, ratologists have learned much about the rat's habits and have put this information to work. It has been found, for example, that the most effective rat kills are made in winter, when breeding is low; populations effectively attacked in winter take a full year to return to normal, twice as long as rat populations attacked in the spring. Another factor is the new poisons that make spectacular kills possible: U.S. Department of Agriculture exterminators, for example, once killed 8.5 million rats in a single three-state operation, using 400,000 traps, carloads of poisoned grain and over a million pounds of sausages treated with poison. Equally important is the vast amount of money the federal government has poured into the war against rats, though much more is needed. All of these factors are crucial, but most important of all, knowledgeable rat killers agree, is the education of people to the dangers of rats and the importance of sanitation in eliminating or controlling them. "Rats cannot thrive without the presence and cooperation of man and any rat problem is basically a people problem," says one official. "We've gone a long way toward educating the public, but we have miles more to go before we sleep."

Around the world the problem is basically the same—the rat problem is a people problem, we have met the enemy and he is us—from Italy, where the rats are killed, appropriately enough, with poisoned parmesan cheese, to India, where religious obstacles to the killing of rats are being overcome. Only in Denmark and a few other countries have the problems of ignorance and apathy been completely overcome.

This is not to say that great efforts are not being made. In the Philippines, where a subspecies of the black rat has populations of *one to a square yard* in many places, and rats often devour a farmer's entire crop, the Rodent Research Center, a joint project of the Philippine Government and the U.S. Agency for International Development (AID), is trying to solve the horrendous rat problem. Everything from the latest poisons to high-voltage electric fences have been tried in the fields, the farmers organize rat hunts on which it is not uncommon for a group to kill one thousand rats, and rat meat has even been introduced as a food (see chapter 13). It remains only for flamethrowers to be brought in to blast the rats out of their burrows, as has indeed been done elsewhere.

So numerous are the big, shaggy, bandicoot rats (*Bandicota bengalensis*) in India's cities that they stand side by side with the cats in alleyways, gourmandizing garbage airmailed out the windows. The cats are afraid to attack them and seem to prefer the garbage, anyway. Rodent-control workers do their best to remove the garbage and harborage every day, but as soon as it is cleared away more rains down. At least 4,000 rats are killed daily in Bombay alone, one half of these by the new poisons and 2,000 each night by nearly 100 "night killers," as these exterminators are called, who walk through the dark city stunning rats with their flashlights and clubbing them to death. Plague-ridden India (a bubonic epidemic raged there as late as 1952) has the farthest to go in rat control, but public education seems to be making progress against the centuries-old Hindu reluctance to kill any animal. Hindus especially revere the rat because their elephant-headed god Ganesha, the symbol of prosperity, is traditionally represented as being transported by rats. At the temple of the Hindu goddess Bhagwati Karniji in Deshnoke, Rajasthan, rats are definitely not "rodentia non grata," they are welcome, as are all living creatures, and they are allowed to roam about freely, even being fed to the tune of some $3,500 a year. Nevertheless, the Indian government has managed to launch successful campaigns against rats in many villages by showing graphic films of how the rodents spread disease and devour or spoil much needed food. A national "rat ministry" has been proposed, and India now has a national rodent control program that is receiving great support from the people, the country having progressed far beyond the point where pythons and mongooses were the main antirat controls. In neighboring Burma, where Buddhists vow to abstain from taking life, the government took a twofold approach in encouraging farmers to kill rats that were destroying the groundnut crop. Officials advised that any religious *Akutho* (demerit) earned by such killing would come to the government, not the individual farmer. The government also persuaded the farmers that the income they made from the sale of the salvaged groundnut crop could be used to help acquire enough *Kutho* (merit) to more than offset any *Akutho* they might have earned by killing rats.

Various poisons constitute the most effective weapons in the modern war against rats. The oldest rodenticide is red squill, an ancient rat poison known to rat killers since at least 1500 B.C. and still widely used. Red squill (*Urginea* or *Scilla maritima*) is a flowering plant native to countries bordering the Mediterranean. Its large bulbous roots, weighing up to four pounds, are cut into slices, dried in the sun and later ground into various preparations, including, at one time or another, a digitalis-like heart medicine, an expectorant, and a diuretic. Today red squill is used mainly as a rat poison in powder or liquid form. Its bitter taste and natural emetic action contributes to its safe use as a rodenticide. Rats, like most rodents, and unlike humans and most other animals, are unable to vomit and are not protected by red squill's emetic quality. Red squill kills by paralyzing the rat's heart. It is effective enough in a single, thorough treatment to reduce a Norway rat population (black rats don't readily accept the bait) but its reacceptance by rats surviving the initial baiting is poor. This single-dose poison is still considered one of the most effective and safest rodenticides, and rats seem to love it mixed with hamburger meat.

Another plant once used, if obliquely, against rats is rue. In the Middle Ages people used to hang branches of rue (herb o'grace) at the sides of windows in their houses to protect against entry of the plague. Although rats were not suspected as carriers of the Black Death, this practice must have helped, because rats do hate rue.

While rue is no longer used as an antirat weapon, many single-dose poisons besides squill are still employed. Strychnine, an old standby, kills rodents within a half hour by paralyzing the central nervous system, but like arsenic (ratsbane), barium carbonate, phosphorus paste, and thallium sulfate, other old rat poisons that no longer are in general use, it is far too dangerous where there are children or pets. Zinc phosphide is a better rodenticide, also killing by heart paralysis, but warning off humans by its definite garliclike odor and strong taste.

A single-dose rat poison essentially nontoxic to man and most effective is the odorless norbormide, which kills by constricting a rat's blood vessels. ANTU

(Alpha-naphthylthisurea) is a slow-acting rodenticide that's safe as far as man is concerned, but it has a bad safety record with dogs, cats, and other domesticated animals.

The three most deadly *and dangerous* rat poisons are Vacor, 1080 (Sodium fluoroacetate), and 1081 (Fluoroacetamide). Human deaths have been reported in Korea and the United States from the eating of Vacor rat bait. After California alone reported nine cases of adult poisoning by Vacor in 1980, the Environmental Protection Agency banned its use. Vacor, however, is a mild single-dose rat poison compared to 1080 and 1081, almost identical poisons of which 1081 is a shade safer and easier to manufacture. Both 1080 and 1081 are among the ten most toxic substances in the world, causing death by paralysis of the heart and central nervous system within one to eight hours. They are extremely dangerous to humans and other animals, having killed dogs and cats after they ate rats poisoned by them. When three children in Durant, Oklahoma, died from eating wafers soaked with 1081, which they had found in a pest control operator's truck, the Environmental Protection Agency restricted the use of the poison compound to sewers and allowed it to be sold only to licensed commercial exterminators. The use of 1080 is also restricted almost entirely to bonded professional operators. Recently, Fort Worth authorities warned residents to be on the lookout for a small canister of 1080 taken by burglars who looted an exterminating company, fearing that the poison could get into the city's water supply by accident. They had reason to be concerned, for a dose of 1080 the size of a matchhead can kill an adult. Regulations are necessarily very strict in this area. In fact, a pink-dyed rat poison was taken off the market in 1981 because the dye was found to contain a substance that caused cancer in lab rats!

Single-dose rat poisons are on the whole much more dangerous and far less effective than the new slow-acting anticoagulants, as the comprehensive chart of rat poisons in Appendix III clearly shows. A revolution in rat killing occurred in 1950, when anticoagulants were developed at the University of Wisconsin by scientists investigating why cattle often hemorrhage and die after eating spoiled hay. The scientists isolated a chemical called dicoumarol that invariably killed the white laboratory rats they were using in their experiments. From this discovery came the first anticoagulant compound, warfarin—a deadly rat poison developed, ironically enough, by man and rats.

Less than 1/500th of an ounce of warfarin is enough to make a rat die of internal bleeding, which is the way anticoagulants work, the rats literally bleeding to death. Anticoagulants are particularly effective against rats because they are acceptable to the rodents in bait and do not cause bait shyness. As noted, rats are among the most paranoid of creatures and have the ability to detect miniscule amounts of poison in food. Furthermore, they are so bait shy after eating poisoned food that they would rather starve than eat the same food again; sometimes they will even urinate or defecate on the poison bait to warn other

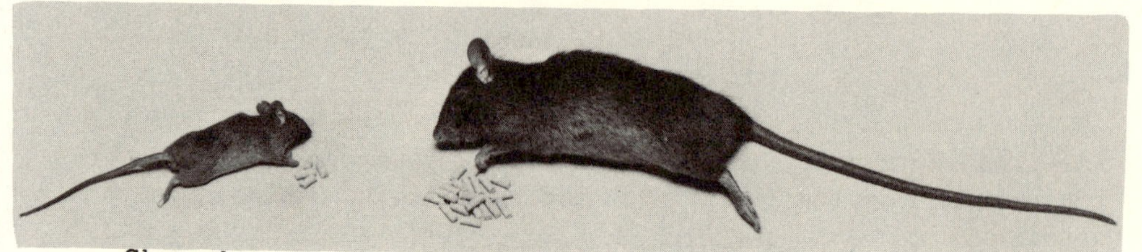
Shown here are the small amounts of modern poisons that are needed to kill a mouse or rat with a single feeding. *(Bell Laboratories, Inc.)*

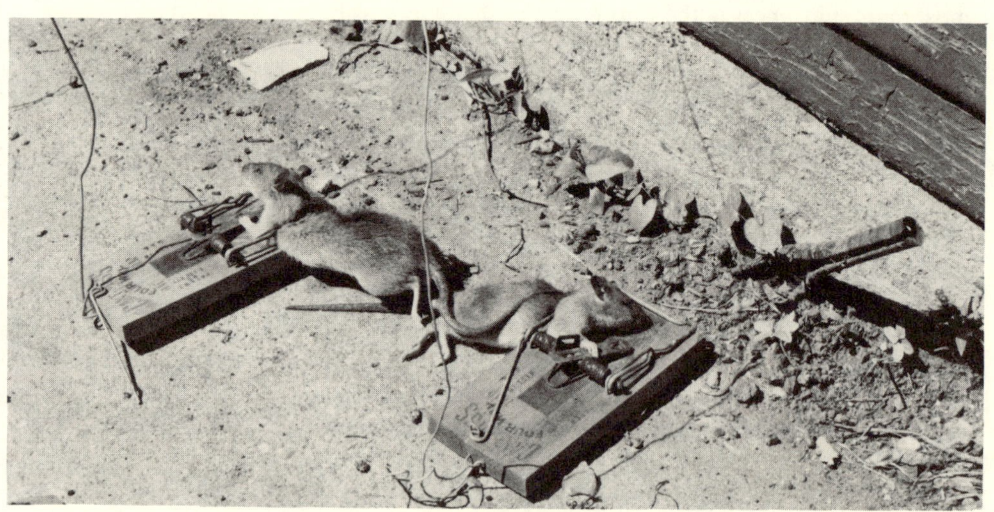
Rats killed in "backbreak" traps *(Chicago Rat Control)*

Gassing rat burrows under a corn crib with calcium cyanide *(U.S. Bureau of Sports Fisheries and Wildlife)*

"Before" and "after" photos showing removal of conditions causing rat infestation in a residential area *(Chicago Rodent Control)*

Chicago rat-control crew baiting sewers *(Chicago Rat Control)*

members of the pack. In *Consuming Passions,* Peter Farb explains it this way: "Rats that have recovered from a single episode of poisoning usually reject, for the rest of their lives, any food with the same taste as what made them sick. Bait shyness thus appears to be a special way of learning long-term avoidance from a single bad experience. This is quite different from classic Pavlovian conditioning in which several pairings of stimuli are necessary to produce a conditioned response, and in which the response is gradually extinguished unless it is reinforced from time to time by repetition. The kind of learning seen in bait shyness must have evolved as an adaptation that saved mammals from having to carry around extra mental baggage to cope with the same experience each time it occurred."

In any case, anticoagulants never cause bait shyness. For this reason and because they are very effective, easy to apply, stable, economical, and relatively safe to use around man and his animals, they have been the rat poisons of choice since 1950. They include warfarin, (which was also used as a heart drug to treat President Eisenhower), Fumarin, Pival, diphacinone, and PMP, working as food baits or as a tracking powder or dust that adheres to the feet and body hair of the rat and is ingested as the animal grooms itself. Anticoagulants must be eaten several times before the rat dies, this multiple-dose feature providing a margin of safety for a child who might eat a single large portion of the bait. Even when weakened from anticoagulant intestinal bleeding, rats don't appear to associate their condition with the food supply, returning again and again to feed on the anticoagulant bait.

When the first "super rat" came to light on a Scottish farm in 1958, the rat-killing fraternity feared the worst. So-called super rats resistant to anticoagulants were reported in a number of other European countries and by 1971 were found in a rural area near Raleigh, North Carolina. Tests made at the Environmental Studies Center of Bowling Green State University and at the Rodent Control Evaluation Laboratory of the New York State Health Department on over 1200 Norway rats showed that more than 5 percent of the animals tested were resistant to warfarin anticoagulant poison in 25 U.S. cities. Some cities claimed far more super rats. In Chicago, for example, about 75 percent of the Norway rats were super rats and in New York City 35 percent fit that description, a figure that promised to increase to 80 percent in three to five years. The resistant rats have been explained in the following way: anticoagulants kill rats by preventing their blood from clotting, causing internal bleeding, the use of vitamin K in the liver being the process obstructed by the anticoagulant. In normal rats, anticoagulants prevent vitamin K conversion and subsequent formation of blood-clotting proteins so that the animals die of internal bleeding. On the other hand, resistant or super rats can always chemically convert vitamin K, enabling them to survive up to 100 times the dosage of anticoagulant poisons that would kill a normal rat. The anticoagulants kill off the most susceptible of the rats and there is interbreeding of the naturally resistant survivors until a significant percentage of the population survives after feeding on anticoagulants—all in keeping with Darwin's concepts of great variability among individuals within an overpopulated species, a struggle for existence, and the survival of the fittest.

But new substitutes have been found that succeed where anticoagulants fail. One recommendation is that anticoagulants still be used and be supplemented every six to twelve months with single-dose poisons such as red squill or zinc phosphide to kill off the resistant super rats. Other experts feel that "super rat" is a misleading term, anyway, pointing out that when 5 percent of any rat population develops immunity to a poison, exterminators should simply change poisons. This certainly worked in Chicago, where in 1981–82 the new "second generation" anticoagulants Talon and Chloro-phacinone (Rat Guard II) were introduced against the resistant rats. "As the result of both these new rodenticides being made available," advises First Assistant Commissioner Madison L. Brown II of the Chicago Department of Streets and Sanitation, "I can safely say Chicago's rat problem has been reduced 70 percent during the past 18 months, with no reports of accidental poisonings or adverse effects in our use of the rodenticides."

How a rat is served an effective poison is often as important as the poison itself. Though Italian exterminators serve their poison with parmesan cheese and it has been mixed with Egg Foo Young in Chinatowns, the rat's favorite foods (not a cheese among them!) include fresh, frozen, or canned meats; fish or pet foods; bacon; yellow corn meal; cracked corn; hulled oats; rolled oats; apple; sweet potato; melons; tomato; molasses; peanut butter (first used as a substitute for rationed or hard-to-get cheese during World War II), and all kinds of nutmeats. The perfect gourmet dinner for a rat might then be:

APPETIZER —Honeydew melon or cantaloupe
SALAD —Tomatoes and cornmeal with peanut oil dressing
ENTREE —Fish-oil-fried hamburger wrapped with bacon
VEGETABLE—Sweet potato-peanut butter soufflé
DESSERT —Sliced apples and walnuts in molasses sauce

Such rat baits, however good, must of course be liberally laced with poison, placed properly along the greasy rat runways and be renewed regularly with fresh offerings. Needless to say, precautions must be taken to protect children and pets or livestock against accidentally eating the poisons. Such accidents are common enough; in an exterminating campaign during a subway excavation near New York City's Central Park Zoo, for example, the poisoned rats staggered into the zoo and were eaten by animals there. Many of the animals that ate the rats died of poisoning.

Poisons aren't the only modern weapon in the war against rats. As noted, scientists are still experimenting with chemical and radiation devices for sterilizing the rodents. Rats are also killed by ultrasonics. The advice that if a man can make a better mousetrap the world will beat a path to his door certainly proved true in the case of Bob Brown, a musician from Pine Valley, California, who in 1972 tangled some wires while fixing his electric guitar and noticed that the sound produced sent local rats scurrying off. Mr. Brown incorporated his discovery into a "rat repellent box" called "the Ecology Machine." More than 50,000 of these have been sold at from $100 to $500 each and Mr. Brown is now a millionaire. The frequency of the note sent out by the device is far beyond the 20 kilohertz that can be heard by the human ear. The inventor claims that any rats within its range become so disturbed that "their normal life styles are dramatically affected. They either flee or become stunned and act in an erratic fashion, not eating, drinking, or mating. Many simply curl up and cease to function and they dehydrate. There are no harmful effects at all on humans or farm animals."

Not long ago, a similar device was reported in the pages of the *New Scientist,* "a new electrical device that clears a building of rodents by emitting an ultrasonic screech. The Italian manufacturer, Erti Electronics, claims that mice and rats simply cannot stand the noise and beat a hasty retreat. . . . The device is left running all the time and . . . clears 250 square meters of any infected building within a week."

Advertising hype aside, the studies that have been made on electronic antirat devices have often yielded completely opposite results. In 1976, four commercial ultrasonic noise generators were tested at the Rentokil Unlimited laboratories in England to assess their effects "on the feeding behavior of colonies of wild Norway rats in simulated field conditions." The commercial units were placed about 20 inches from the rats' accustomed feeding sites, the study reported. "Although there were initial reductions in food consumption, the reductions were either insignificant or short-lived. The results are in line with previously published studies. Ultrasonic devices currently on the market are largely ineffective in repelling rat populations. In addition, the problem of placing such units so as to cover large areas (high frequency sound dissipates rapidly and does not penetrate solid objects) and the capacity of rats to become habituated to the noise confirms that ultrasonic devices are unlikely to plan an effective role in rodent control."

But according to a study made at the Univeristy of Detroit in 1977, ultrasonics can work against rats. "A large-scale central system to effectively control rats in warehouses and other similar environments has been developed," this study reports. "The system employs high-powered ultrasonic noise which drives rats out of the noise-covered area and protects against further infestation. The ultrasonic noise emitted from the numerous transducers does not disturb humans in the area, for it is higher in frequency than the normal hearing threshold. The central system approach, utilizing one amplifier, signal source and a distributed network of transducers, eliminates duplicity of electronics and this method leads to low cost and high reliability."

Most investigators feel that ultrasound has thus far been of limited use in rat control, reducing but not eliminating the number of rodents in an area. D. J. E. Broads summarized much of the research in this field when he wrote: "Ultrasound will not drive rodents from building areas; will not keep them from their usual food supplies, and cannot be generated intensely enough to kill rodents in their colonies. Ultrasound has several disadvantages: It is expensive, it is directional and produces sound shadows where rodents are not affected, and its intensity is rapidly diminished by air and is thus of very limited range." Nevertheless, perfected ultrasonics, and biosonics (recordings of rodent distress sounds) may prove an important part of future rat-control programs, as may the ultrasonic contraception device previously mentioned.

Rats are sometimes eliminated by flooding their burrows (in regions with tight clay soils), either drowning them or forcing them out into the open where they can be clubbed to death, or, more commonly, by gassing and trapping. Gassing, often used in cities against huge populations of rats, is done with calcium cyanide, carbon bisulfide, methyl bromide, and carbon dioxide. Great care has to be taken when employing these toxic gases and burrows are never gassed less than twenty feet from a building because the gases may seep into the structure and kill people or pets.

The determined rat killer can walk into any hardware store and select rattraps ranging from snap traps, steel traps, and box traps, to elaborate wire cages with funnel-shaped entrances. One Rube Goldberg device on the shelves seizes a rat in steel jaws, promptly electrocuting it and dumping the offender into a plastic bag. Then there are popular shinglelike devices coated with glue in which the rat gets stuck when he walks across it, though rats have been known to bite off their legs rather than remain glued to the shingle. There is a story of one rat caught in such a trap by its tail. The ladies who set the trap discovered the rat and went to find a weapon to deliver the coup de grace. When they returned they found a second rat on the scene who bit off the trapped rat's tail and freed it.

As if the rats themselves were not enough, rat exterminators must also deal with the control of ectoparasites of rats such as lice, mites, ticks, and the fleas that cause bubonic plague. Modern methods call for the careful spraying of all runways, burrow entrances, and nests with carbaryl (Sevin), or ôther insecticides. In plague areas bait stations can be set up so that wild rodents get a harmless dusting of flea-killing insecticide upon entering a trap. As for dead rodent odors, another rat-related problem, these are usually masked when the rodent cannot be removed, and products such as Bactine, Dutrol, isobornyl acetate, Neutroleum Alpha, quarternary ammonium compounds, Styamine 1622, Zephiran chloride, oil of pine, oil of peppermint, oil of wintergreen, formalin, anise, or activated charcoal are sprayed as aerosols as close to the dead rodents as possible.

The main trouble with modern poisons, traps, and other rat-killing methods is that as soon as they effectively reduce the rat population, the prolific animal's staggering reproduction rate promptly replenishes its numbers. Even if poisoning is undertaken during the winter, when the rats' reproductive activity is lowest, the rat population returns to the same level within a year from the time poisoning is stopped. "Poisons or traps," says rat-control expert Dr. David E. Davis, "merely make space for more rats to grow." At best, poisoning rats is only one prong of a three-pronged attack that must be taken if rats are to be controlled in an area of any significant size, the other two being (1) environmental sanitation to eliminate food and harborage for rats, and (2) effective ratproofing.

Rodentologists formulated this battle plan in the war against rats during World War II, when a long-term study of rat-control methods was begun in Baltimore because military intelligence feared that the enemy would secretly unleash a campaign of ratborne germ warfare in America. The germ warfare was never launched, but the study led scientists, under the direction of John Hopkins University psychobiologist Dr. Curt P. Richter, to formulate the new battle plan of environmental manipulation that is today used in all urban wars on rats. It was found that the effect of predators, including man and his traps and poisons, was only temporary in reducing rat populations. Competition, whether between members of the same species or between two or more species, was a far more important factor in permanently limiting the number of rats in an area. This finding led to the basic rat-fighting axiom: *Man should so change the environment through better sanitation as to cause increased competition and predation, thereby lowering the capacity of the environment to support rats.* "If you don't feed them, you won't breed them." Deny the rat food, water, and living space and the rats will compete violently for what little remains, their population dwindling as a result and becoming easier prey for all predators, including cannibals among them, and man.

Altering the environment is possible and has been done, though it isn't easy and always depends upon the cooperation of the community. "Sanitation is basically a matter of human behavior," Chicago Rodent Control Assistant Director Terence Howard concludes, and he might have added that that human behavior toward rats sometimes can be quite odd—from the Calcutta office workers who sit in Curzon Park at noon feeding wild rats the scraps of their lunches, to the well-dressed lady New York pest-control workers found feeding her "pet rats" on Park Avenue not long ago. But rats can be controlled if one follows the proper rat riddance procedures developed by expert exterminators over the years (see Appendix IV).

British novelist Samuel Butler wrote about the mental process of an expert rat trapper:

> Dunkett found all his traps fail one after another, and was in such despair at the way the corn got eaten that he resolved to invent a rat trap. He began by putting himself as nearly as possible in the rat's place.
>
> "Is there anything," he asked himself, "in which, if I were a rat, I should have such complete confidence that I could not suspect it without suspecting everything in the world and being unable henceforth to move fearlessly in any direction?"
>
> He pondered for a while and had no answer, till one night the room seemed to become full of light, and he heard a voice from Heaven saying, "Drainpipes." Then he saw his way. To suspect a common drainpipe would be to cease to be a rat.

Thinking like a rat is indeed the only way to beat the wily rodents, experienced ratcatchers say, and this method has certainly worked for the urban rat-control program in the United States. But it has been suggested that until elected officials and other politicians *stop* thinking like rats and deserting the ship of public health for one foolish reason or another, rat extermination on a national or global level is not a real possibility in the foreseeable future. The U.S. program, administered through the Centers for Disease Control of the Department of Health and Human Services under Section 317 of the Public Health Service Act, has already offered conclusive proof that rats *can* be effectively controlled. Since 1969 the program has reduced rat infestations to a level at which they no longer constitute a major problem in 36 of the 68 participating communities. With a Federal budget totalling only $165 million over the past eleven years and matching funds of $193 million contributed by state and local governments, Urban Rat Control has enabled more than 7 million people to finally live in virtually rat-free areas at a cost of just $50 a person, a total of 59,391 blocks in cities across the country now rat free, with 20,526 more blocks targeted for the near future. However, in just the United States, fully two-thirds of communities with significant rat problems must still be reached, and Dr. Vernon Houk, physician director of the federal program, has gone on record bluntly calling federal funding "peanuts," considering the size of the problem remaining. Yet, at a time when *more* money is needed, the present administration is proposing massive cuts in the public health sector of up to half a billion dollars in a reorganization that would eliminate the authority of the nation's top public health officials and not only prevent further gains in the war on rats, but cause the loss of ground already gained.

This shortsighted tactic is nothing new or peculiar to one administration or nation, in the present, or throughout history. Though we have met the enemy in the war against rats and learned that "he is us," as Pogo observed, the lesson hasn't been learned well enough. If the rat is the Nazi of nature, the world, to which some refuse to believe we belong, is about as well-prepared for its future invasions as Belgium was against Hitler's blitzkrieg. Like George Orwell in *Homage to Catalonia* we may wake up one fine morning to find we have been sleeping with rats. Ironically, there will be catastrophic ratborne epidemics, we will continue paying a rat tax of billions of dollars a year in damaged property, rats will go on biting babies, living better than some people do, and killing people, until people stop acting like them in one way or another—in the lower depths of the streets, where the environment is degraded by rich and poor alike, and in the rarified air of high government offices, where human decency is degraded. Only when we—especially the Pied Pipers among us—stop emulating rats, will it be possible that humankind may outlast ratkind. . . .

*It is winter, time for a massive rat kill but the city has no money, there are heavy snowfalls, the few pest control workers remaining due to budget cuts have been diverted to snow removal duties. Yet people don't stop trashing, garbage usually removed accumulates, the city's eight million rats feed and feed and feed. . . . By spring the rats number eleven million . . . another budget cut the next winter, in a year there are 20 million rats, three to every person . . . rat bites soar in number . . . insidious diseases spread . . . the rats are of plague proportions . . . they boldly claim the streets and city . . .*

# VII

# NOW, MUSE, LET'S SING OF RATS

## The Rat in Literature, Art, and Folklore

Some phony Pied Pipers may be leading the children to Armageddon, but the real Pied Piper is alive and well and doing his job in Hameln (the real Hamelin). He is no myth, but a very substantial German *Rattenfänger* with black hair, a strong set to his mouth, and gentle eyes betraying no malice toward the townspeople who cheated his famous predecessor, considered the greatest of ratcatchers past and present.

In an interview with *National Geographic* reporter Thomas Y. Canby not long ago, the Pied Piper did indicate that he, too, like his ancestor, has his money problems. He isn't getting the money he deserves. "I would earn twice as much [catching rats] in any other city this size," he complained. The stocky, thick-browed *Schädlingsbekämpfermeister* (master pest fighter) is Hameln's modern-day counterpart of the Pied Piper. The local papers even call him that. Wilhelm Klimasch patrols the back alleys and sewers of Hameln keeping the legendary north German town practically rat free, or *praktisch rattenfrei,* as an official document in the town hall proudly proclaims, but Hameln's stingy citizens still don't want to pay the piper—not what he's worth, in any case. Hameln, which still looks like a storybook medieval town with its ancient churches and half-timber houses, goes to great lengths to identify itself with the legendary Hamelin of old. Tradition even demands that no music may be played on Bunger Street, through which the first Pied Piper piped his child-victims to their deaths. Yet twice a day in an old stone *Rathaus* with murals of the tale on the walls, a glockenspiel re-enacts the story of the ratcatcher who was cheated of his wages and had his revenge. From bakery windows, ingenious bewhiskered pastry rats stare out at tourists and local grog shops sell wine and beer in bottles shaped like rats. Whether the burghers inherit stinginess from their ancestors or are merely underpaying Wilhelm Klimasch to exploit the mythical tale more is a matter of opinion. Klimasch, however, remains because there is no more prestigious position on earth for a ratcatcher. So far he has not taken any revenge, judging by all the children on the streets.

"Might you consider luring away the children sometime in the future?" he was asked as he poked his head out of a manhole.

"I would be tempted," he said, with just the trace of a smile. "But I cannot play the pipe!"

In any event, of all the tales of rats and men none is better known than *The Pied Piper of Hamelin,* a legend which English poet Robert Browning retold in a poem he wrote for little Willie Macready, the son of one of his friends, and never intended to publish.

"Rats!" Browning wrote:

> They fought the dogs and killed the cats
> And bit the babies in the cradles,
> And ate the cheeses out of the vats,
> And licked the soup from the cooks' own ladles . . .

Until the Pied Piper came on the scene:

> To blow the pipe his lips he wrinkled
> And green and blue his sharp eyes twinkled . . .
> And ere three shrill notes the pipe uttered . . .
> Out of the houses the rats came trembling—
> Great rats, small rats, lean rats, brawny rats,
> Brown rats, black rats, gray rats, tawny rats . . .
> And step for step they followed dancing
> Until they came to the river Weser.

Browning's poem with lines like "[rats] shrieking and squeaking / In fifty different sharps and flats." or "Anything like the sound of a rat / Makes my heart go pit-a-pat," made the old tale of the Pied Piper familiar to readers everywhere. Reputedly based on fact, the legend has it that Hamelin (Hameln), Germany, became infested with rats in 1284 and contracted with a strange Pied Piper, so named because he dressed in a two-color, or pied, costume, to rid the town of them. This Pied Piper drowned the rats, who followed the mysterious sound of his pipe to the Weser River. But when the village refused to pay him his 1,000 guilders, he returned the next St. John's Day, and on that June 26th piped away all the town's 130 children, only a lame child and a blind child not disappearing forever into the cave where he led them on Koppelberg Hill. (A later legend held that he led them *around* the mountain to found a town of children in Transylvania.)

Since the original story is rarely told, here is the very first English version, from Richard Versigan's *Restitution of Decayed Intelligence* (1634):

> There came into the town of Hemel an odd kind of companion, who, for the fantastical coat which he wore being wrought with sundry colours, was called the Pied Piper. This fellow, forsooth, offered the townsmen, for a certain sum of money, to rid the town of all the rats that were in it (for at that time the burghers were with that vermin greatly annoyed). The accord, in fine, being made, the Pied Piper, with a shrill pipe, went through all the streets, and forthwith the rats came all running out of the houses in great numbers after him; all which he led into the river of Weaser, and therein drowned them. This done, and not one rat more perceived to be left in the town, he afterward came to demand his reward according to his bargain; but... [the townsmen] offered him far less than he looked for. He, therewith discontented, said he would have his full recompense according to his bargain; but they utterly denied to give it to him. He threatened them with revenge; they bade him do his worst, whereupon he betakes him again to his pipe, and going through the streets as before, was followed by a number of boys out of one of the gates of the city, and coming to a little hill, there opened in the side thereof a wide hole, into the which himself and all the children did enter; and being entered, the hill did close up again, and became as before. A boy, that, being lame, came somewhat lagging behind the rest, seeing this that happened, returned presently back, and told what he had seen; forthwith began great lamentation among the parents for their children, and the men were sent out with all diligence, both by land and by water, to inquire if aught could be heard of them; but with all the inquiry they could possibly use, nothing more than is aforesaid could of them be understood.

According to the Reverend Sabine Baring-Gould (the author of "Onward Christian Soldiers"), who discusses the legend in his *Curious Myths of the Middle Ages* (1896), the real Piper's name was Bunting. But scholars can't agree on the origin of the legend. Some historians say it was based on an actual happening, and some claim that the real Piper worked for an official who organized a group of young men to settle in Moravia in 1284. Others say the tale is really a plague story in disguise, that Hameln refused to pay a ratcatcher during the plague of 1348 and the town's children died of plague, coincidentally, after cleaning up the rats he had killed in town. Still other scholars connect the story with the Children's Crusade in the Middle Ages, when armies of children were gathered in Europe to march off under a charismatic German youth named Nicholas to fight the infidels in the Holy Land, twenty thousand dying on the long march and many more sold into slavery when captured.

Another Pied Piper story involves the fiddler of Brandenberg, and still another takes place over a three-year period in the town of Lorch starring several heroes or villains, the first of whom is an old hermit piper. When Lorch became infested with ants, the hermit piped the vermin into a local lake, where they all drowned. As the inhabitants refused to pay the stipulated price, he led their pigs into the same lake. The next year a charcoal burner cleared Lorch of crickets, and when the locals again reneged, he led their sheep into the lake. In the third year an old man of the mountain piped away all the rats that plagued the town to a watery death, and when the inhabitants refused to pay he charmed all the children of Lorch into the lake.

A medieval account of the Pied Piper of Hamelin transforms the ratcatcher into the devil: "There the devil in human form walked through the streets on the day of Mary Magdalene, blew his pipe and enticed many children, boys and girls, to himself, and led them through the town gate to a hill. When he arrived there, he vanished with the children, of whom there was a very large company, so that no one knew where the children had gone. This was related by a little girl who had followed from afar to her parents, and diligent enquiries were instituted by land and water, and news sent to all places to ascertain if perhaps the children had been stolen and taken away. But no one ever found out where they had gone to."

Scholars have cited the ancient literature of Persia and China as sources for the Pied Piper of Hamelin legend. Parts of the story may have been inspired by Greek and Sanskrit mythology, especially the irresistible lure of the siren's song in Homer's *Odyssey*. Certainly rats were a great problem in medieval times, when bubonic plague spread by rat fleas killed more than 75 million people. In fact, the colorfully dressed professional ratcatchers carrying a pole bearing a flag with a portrait of a rat on it were common in the Middle Ages and the Pied Piper could well have been based on one of their kind. At any rate, the popularity of the intriguing story is based to an extent both on the proverbial moral that one must pay the piper and on the very real problems of rat control which still plague men today. The tenacious legend persisted for more than five centuries, the last version concerning the people of Isfahan, who toward the end of the seventeenth century were tormented with rats when a little dwarf named Giouf, not two feet high, promised to free the city of the rodents in an hour if he were paid a certain amount of money. The terms were agreed upon, and Giouf piped every rat into the Zenderou River, where all drowned. When the dwarf demanded payment, the people gave him counterfeit coins, refusing to take them back, and Giouf took his revenge on the people and their children.

To this day no one knows what kind of pipe the Pied Piper used to lure away Hamelin's children. It could have been a flute or a piccolo, a syrinx, an Egyptian nay, or any odd or forgotten member of the wind instrument family embraced by the generic name of "pipe." But since rats do hear shrill sounds much better than deep ones (see chapter 6), the choice of a pipe was a good one. The rats may have heard irresistible shrill pipe notes that were beyond the range of human hearing!

By no means is the Pied Piper of Hamelin the first or last legend about rats. "Now, Muse, let's sing of rats" began a manuscript by British author James Grainger entitled *The Sugar Cane,* or so goes the story in Boswell's *Life of Johnson.* Goethe wrote a poem *Der Rattenfänger (The Rat Catcher),* which has been put to music by Schubert, Wolf, Gounod, and Berlioz over the years. Rats also figure in Prokofiev's opera *The Love for Three Oranges* and in several of La Fontaine's fables set to song by the Dutch composer Koumans. Other authors have had an interest in the subject, too, from Shakespeare, who mentions rats at least seven times, notably in *Hamlet* and *Lear,* and Swift, who has giant rats in *Gulliver's Travels,* down to Arthur Conan Doyle, who mentioned the giant rat of Sumatra; and H. G. Wells, who wrote of huge rats in *The Food of the Gods.* Since "the rat that ate the malt that lay in the house that Jack built" to modern horror film classics like *Willard* and *Ben,* the rat has played an important part in literature, usually as a symbol of evil, often serving authors well as a virtual prop in suspense stories. George Orwell's *1984,* Albert Camus' *The Plague,* William Faulkner's *The Reivers* and H. P. Lovecraft's horror stories all feature rats. So does Selma Lagerlöff's *The Wonderful Adventures of Nils* and Kafka's unfinished story "The Burrow." Artists and sculptors have displayed a similar interest in rodents, from early Chinese and Roman days up to the time of Mickey Mouse (see chapter 10). Rembrandt did a number of prints of rats and rat-catchers, as did Cornelis Vischer, Pieter de Bloot, and Jan Steen, among others.

Aside from La Fontaine's fables, the best-known work in which rats are treated sympathetically is Kenneth Graham's charming novel *The Wind in the Willows,* in which both the brown Water Rat and the black Ship Rat are good-natured hardworking fellows. Edgar Allen Poe's story "The Pit and the Pendulum" gives an excellent description of ravenous rats attacking a helpless man: "They pressed—they swarmed upon me in ever accumulating heaps. They writhed upon my throat; their cold lips sought my own; I was stifled by their thronging pressure; disgust, for which the world has no name, swelled my bosom, and chilled, with a heavy claminess, my heart . . ." Yet the rats in Poe's tale became heroes of sorts because they unintentionally free the prisoner by gnawing through his bonds.

Martin Luther called the pope "the king of rats," terming Catholic cardinals a "rabble of rats" and monasteries "rats' nests." From ancient times, the rat has been hated (not surprisingly) and loved (astonishingly) by humans. The Phrygians and Egyptians, as examples of the latter, both deified the rat, and the natives of Bassora, Turkey, and Cambay, India, up until recent times forbade the destruction of rats, believing them to bring good luck. To the ancient Egyptians the rat symbolized utter destruction but also stood for wise judgment, because a rat "always chooses the best bread." The Japanese have long held that the rat is messenger of the god of wealth and that if one gnaws New Year rice cakes there will be a good harvest that year. The Chinese, too, considered the rat to be a symbol of prosperity, but they had an ambivalent attitude toward the animal centuries before the rat was known as a carrier of plague and disease. In China the "Year of the Rat," which falls every twelve years, is considered an unlucky year in which to be born.

All rodents were considered as unclean as pigs by the ancient Jews, and the worshippers of Zoroaster believed that the killing of any water rat was a great service to their god. One story has it that King Sennacherib of Arabia and Assyria marched a great army against Egypt and on the night before the battle "there swarmed upon them" rats or mice that "ate up the quivers and their bows and the handles of their shields" so that on the day of the battle they fled. A similar tale is told of the Roman bucklers at Lanuvium being gnawed by rats, which "presaged ill fortune, and the battle of Marses, fought soon after, confirmed this superstition." Abdera, a seaport in Thrace, is said to have been so overrun with rats that it was abandoned and the inhabitants moved to Macedonia.

The Romans were very superstitious about rats, believing that while black rats meant bad luck, white rats foretold good fortune. We have mentioned how the Roman god Apollo killed a multitude of rats with his far-reaching arrows. A real rat battle was held by the insane emperor Heliogabalus, who "staged a fight between ten thousand mice, one thousand rats and one thousand weasels" in which the rats beat the mice and the weasels beat the rats. Among the rat superstitions held by the Romans was Pliny's advice in *Historia naturalis* that "If a woman be desirous that her infant should be born with black eyes, let her eat a rat while she goes with child." But then Pliny also believed incredible things about the fruitfulness of rats. "It is said that they engender by licking," he wrote, "without any other kind of copulation, and that one of them hath given birth to six score at a time, and also that in Persia there have been young ones found with young even in the belly of the pregnant one," which is a good comment on the rat's amazing fecundity.

During the Middle Ages rats were believed to be creatures of evil, the lapdogs of the devil, even though they were not associated with the deadly plague that they carried. Both the devil and witches were said to change into the form of rats and mice so that they could sneak about and pry into human affairs without being noticed. Saint Gertrude was besought by the bishops of the early Catholic Church to protect against devil rats and mice. The Church went so far as to place a ban on rats on several occasions, notably in 1478 by the Bishop of Berne, in 1516 by the Bishop of Troyer, and in 1541 by the Bishop of Lausanne. From pulpits all over Europe, rats were ordered to leave the land, and, as we've seen, the Irish tried to rhyme them to death.

Another prevalent notion regarding rats, and one still held today, was that they had a presentiment of coming evil and always deserted a ship about to be wrecked, or a house about to be flooded or burned. As late as 1854 a Scottish newspaper reported that the night *before* a town mill was burned to the ground, the rats that inhabited it were seen migrating in a body to a neighboring field. That rats are the first to desert a sinking ship can be explained by the fact that water begins entering a sinking ship below decks where they make their homes. An old American proverb instructed "When the water reaches the upper deck, follow the rats." In both America and England, *to rat* has long meant to forsake a losing side for the stronger party (a *ratfink* would do this today), or to desert one's party, an allusion to rats forsaking unseaworthy ships. Swift wrote in his "Epistle to Mr. Nugent":

> Averting . . .
> The cup of sorrow from their lips.
> They fly like rats from sinking ships.

The air and especially the soil in certain areas were deadly to rats, credulous ancient authors believed; therefore, rats could not live in certain districts, which would always be rat-free! In John Sinclair's *Statistical Account of Scotland* (1794), there is a story about this belief carried to its most absurd extreme. "From a prevailing opinion," writes the author, "the soil of Roseneath parish is hostile to the rat. Some years ago, a West Indian planter actually carried out to Jamaica several casks of Roseneath earth, with a view to killing the rats that were destroying his sugarcanes. It is said that this had not the desired effect; so we lost a valuable export. Had the experiment succeeded, this would have been a new and profitable trade, but perhaps by that time, the parish of Roseneath might have been no more!"

A medieval legend tells of the German count who raised a tower in the middle of the Rhine and had his men kill with crossbows the crew of any boat that wouldn't pay the toll he charged. The cruel count cornered the wheat market one famine year and profiteered at the expense of the people, but rats pressed by hunger wreaked revenge on him, invading his tower and devouring him along with his horde of grain. This isn't the only German legend about rats eating humans, which we know has happened often in modern times (see chapter 1). The cruel Bishop Adolf of Cologne was supposedly devoured by rats in 1112,

The Pied Piper legend may have been based on a wandering ratcatcher like this one depicted by Cornelius Vischer c. 1635. He is a walking advertisement for his trade. *(Chambers's Book of Days)*

The lion and the rat, a pair that have figured in fables and folklore from Aesop on. *(Leslie's Illustrated)*

(above) A Doré illustration for the La Fontaine fable "The Rat Retired from the World." *(N.Y. Public Library Picture Collection)*

(below) A rat tries to get the prize from almost under the cat's whiskers in an illustration for one of the countless cat and mouse (or rat) tales that have appeared over the centuries. *(Harper's Monthly Magazine)*

In this nineteenth-century Doré illustration, the rats meet to decide which of them will bell the cat. *(N.Y. Public Library Picture Collection)*

This humorous old print, entitled "Rats on Board: A Midnight Fantasy," depicts the trials of a sea captain with rats aboard his ship, c. 1880. *(N.Y. Public Library Picture Collection)*

and Freiherr von Guttingen, who "burned all the poor in a barn during a famine in order to save food," was eaten "clean to the bone" by rats in his castle. That rats were really common and numerous enough to eat people is shown in an English *State of Prisons* report about a jail cell in Knaresborough Gaol: "The cell is under the hall, of difficult access, the door about four feet from the ground. Only one room about twelve feet square, earth floor, no fireplace, very offensive; a common sewer from the town running through it uncovered. I was informed that an officer confined here took in with him a dog to defend him from the rats; but the dog was soon eaten and the prisoner's face much disfigured by them."

Alexander Selkirk, the real-life model for Defoe's Robinson Crusoe, had his trouble with rats on the island of Juan Fernandez in the South Seas. "His habitation was extremely pestered with rats, which gnawed his clothes and feet when sleeping," Richard Steele wrote in an English newspaper. "To defend himself against them he fed and tamed numbers of young kitlings, who lay about his bed and preserved him from the enemy." Both the rats and the cats had been left on the island by mariners before him. To this day, evidence of the rats that Europeans introduced exists in the South Seas. On Tahiti, even the cats, landed by Captain Cook to counteract the rats, seem to have done little good, for every palm tree is encircled with metal to prevent the local rats from getting the coconuts.

In the South Seas kingdom of Tonga, which Captain Cook visited, the folktale is still told of the rat that hitched a ride across the pond on the back of a turtle. On reaching shore he laughed scornfully, "See what I have left upon your back, o foolish trusting turtle!" The turtle, however, got his revenge the next time the rat hitched a ride, submerging and drowning him. The story shows little knowledge of rats, as any self-respecting rat would have swum to shore.

Rats figure in the journals of many great sailors like Captain Cook, and few admirals of the ocean have anything good to say about these cowards that are the first to desert a sinking ship. In *Sailing Alone Around the World* (1898) Captain Joshua Slocum, the first person to circumnavigate the earth alone in a ship, pointed out one superstition endearing the rat to mariners, but could only share his ship with rats within limits. "It is, according to tradition, a most reassuring sign to find rats coming to a ship," he wrote, "and I had a mind to abide the knowing one of Rodriguiz [Island]; but a breach of discipline decided the matter against him. While I slept one night, my ship sailing on, he undertook to walk over me, beginning at the crown of my head, concerning which I am always sensitive. I sleep lightly. Before his impertinence had got him even to my nose I cried 'Rat!' had him by the tail and threw him out of the companionway into the sea."

Dr. Jean Charcot, who explored the Antarctic in the *Why Not?* in 1909, became so attached to a ship rat that he almost made a pet of it. "The ship's rat," he wrote in his diary, "the only one since his companion committed suicide by falling through one of the scuppers, after having given no signs of life for two months has again given proof of his existence by eating two birds [from the scientific collection]... It is sad that he is spoiling our collections thus, for the cats seem to trouble very little about him, and we too could easily have put up with him. I had even a scheme for taming him. How this poor solitary rat must be bored and how much he must regret his choice of a ship."

Some people don't mind living with rats at all. Besides the Chicago and Miami ladies already mentioned, there is a case on record of an eighteenth-century Englishwoman who customarily slept with a rat and a nest of mice, preferring them to any man. Another eccentric Englishwoman, the Countess of Eglintowne, who thrived in the late eighteenth century, customarily dined with rats. She rapped on the dining room wall, whereupon the rodents appeared through a sliding panel to eat the repast she always set out for them, disappearing back into the wall when, tired of their company, she rapped again.

We have mentioned one example of the bizarre fondness for rats exhibited by orthodox Hindus, who venerate not only cows but all animal life. Another Hindu sect has a temple consecrated to rat worship in the city of Deshnoke in Rajasthan, India, where hundreds of pampered rats are daily fed bowlfuls of a mixture of sweets, grain, and a milk called *laddu.* These rats have little fear of humans and scores of them mingle freely with worshippers. "While I was shooting pictures," a *National Geographic* photographer back from India recalled, "the rats gnawed holes in my camera bag and even chewed the insulation from my strobe lights, shorting them out." In Calcutta, whose filthy streets are a veritable concentration camp without barbed wire for unknown numbers of wretched souls, there is actually a "Rat Park," where as reporter William Drummond of the *Los Angeles Times* put it, "dozens of the fat brown creatures play and cavort in broad daylight while human pedestrians pass nonchalantly by."

Rats figured ironically in the story of an unsuccessful attempt by German army officers to take Adolf Hitler's life on July 20, 1944. On that night at 7 o'clock, a German radio broadcast announced the assassination attempt and followed through with martial music in place of a previously scheduled discussion entitled "The Extermination of Rats."

Any collection of rat-extermination stories should also include the tale of the man in North Carolina who blew up his house while trying to perfect a chemical rat killer. Worthy of mention, too, is William Henderson of Chicago, who tried in vain to rid his home of rats by sitting up nights taking potshots at them. The army of rats simply went about their business, occasionally pausing to bare their teeth at him. After several members of his family were bitten, Henderson had to abandon the house, which city officials said would be razed as a public health hazard "if the rats don't eat it first."

The most unusual recent story concerning rats must be that of the "rat-men" who reportedly landed in a UFO near Rio de Janeiro. (You will probably choose to believe it not.) These rat-faced creatures "had huge horrible ears, their mouths were like slits, their skin was sticky and gray, and they somehow communicated without talking to each other," according to Brazilian concert pianist Luli Oswald, 55, and her companion Fauze Mehlen, 25, who reportedly saw a fleet of rat-men-operated UFOs "pop out of the ocean" while driving along the coast early in 1982 and who were "held captive" by them for two hours. Under hypnosis Ms. Oswald claimed that the rat-men carried her into their UFO, "put a tube into my ear," pulled her hair, and examined her friend with "a strange ray of light" that "smelled like sulphur." One of the rat-men did speak to her, in Spanish: "He said they came from Antarctica, that there is a tunnel which goes under the South Pole. That's why they come from the water."

Sigmund Freud's famous Rat Man, readers will remember, didn't come out of a UFO or resemble a rat in any way; he, rather more simply, had an obsessional neurosis about rats. This Rat Man consulted with Dr. Freud some time after he had been told by a fellow army officer of a Chinese torture in which a pot filled with rats was strapped onto a criminal's buttocks and the trapped rats bored their way into his body through his anus. Overwhelmed with the story, the patient began imagining the torture applied to two people very close to him, Freud discovering why he did this in an attempt to cure the man.

There were charges during the Vietnam War that intelligence agents on both sides put cages containing rats around the heads of prisoners, but a closer variation on the sadistic Chinese pot-full-of-rats torture is reportedly employed by African tribes in Northern Uganda, a type of torture used, we are told, by Idi Amin. Here a rat or two is placed on a man's wounded stomach and covered with a pot, which is heated to an unbearable temperature. The rats, with only one way of escape, eat into the wound and through the man's innards until they find a way out, the man forced to watch their progress after several minutes pass and the pot is removed.

The world's biggest hoax about rats (and cats) was probably the prospectus of the Lacon, Illinois, "cat-and-rat ranch" which was sent by an anonymous hoaxster to an Illinois newspaper in 1875 and carried by practically every newspaper in the country, many of the editors fully believing it. Read the prospectus:

> GLORIOUS OPPORTUNITY TO GET RICH: We are starting a cat ranch in Lacon with 100,000 cats. Each cat will average twelve kittens a year. The cat skins will sell for thirty cents each. One hundred men can skin 5,000 cats a day. We figure a daily net profit of over $10,000. Now what shall we feed the cats? We will start a rat ranch next door with 1,000,000 rats. The rats will breed twelve times faster than the cats. So we will have four rats to feed each day to each cat. Now what shall we feed the rats? We will feed the rats the

carcasses of the cats after they have been skinned. Now get this! We feed the rats to the cats and the cats to the rats and get the skins for nothing!

The rat has figured in American literature from earliest times, humorously in Tom Sawyer's swinging his dead rat on a string and in real town names like Rat Gulch, Rat Trap, and Rat Portage; most horribly in stories of which H. P. Lovecraft's tales of Salem graveyards riddled with rat burrows are the epitome or nadir. (His inspiration Algernon Blackwood in England, also had his share of horrible rats.) Don Marquis's "Death of Freddy" piece in his Archy comic strip is far gentler. Recently, Richard Pryor played with a scene-stealing rat in the movie *Some Kind of Hero,* and at least two plays entitled *Rats* have appeared in New York.

There remain many myths about rats, most of them based on some small truth. Rats, for example, can run at a top speed of only about six miles an hour (humans can go more than twice as fast), but they are certainly frenetic running in the treadmills of their cages, sometimes putting in twenty miles a day. This probably suggested the common expression *rat race* for a life style where the results aren't worth all the effort. As noted, the term *dirty rat,* isn't really fair to the rat, but *to fight like a cornered rat* is a saying firmly based in reality—a cornered rat is a vicious animal of which to beware.

Some people really have been able *to smell a rat.* One legèndary underground worker for the New York City transit authority was renowned for smelling out dead rats, the odor of which has ruined many a business. John "Smelly" Kelly, also known as "Sniffy," patrolled the New York City subway system for years, his uncanny sense of smell enabling him to detect everything from gas leaks to the decomposing corpses of poisoned rats that crawled into the walls of stores adjoining the subway to die. Smelly Kelly was one of those rare people who could say with certainty, "I smell a rat." No one, however, can say with certainty how that expression originated, though the allusion may be to a cat smelling a rat while being unable to see it. Terriers and other rat-hunting dogs could also be the source. The expression dates back to about 1780, but long before that "to smell," was used figuratively for "to suspect" or "discern intuitively," as when Shakespeare writes "Do you smell a fault" in *King Lear.* St. Hilarion, the Syrian hermit who died about 371, could allegedly tell a person's vices or virtues by smelling his person or clothing.

Human fear or hatred of rats was most recently expressed in a new video game for children called Sewermania, in which a character "must defuse a timebomb in a mazelike sewer filled with killer rats. . . . With one voice command ('Pick up'), he grabs a shovel. Another order ('Kill') prods the shovel-armed man to bash the rats."

We might finally consider the persistent myth that tells of horrible rats as big as cats, or bigger. Indeed, the movie version of H. G. Well's story "The Food of the Gods" (1975) features rats as big as wolves (made so by trick photography), all led by a giant white rat. These creatures have to be shot, electrocuted, and drowned before the world is rid of them. New York City underground workers often have reported rats as big as cats in the sewers. Never has proof in the form of a dead giant rat, or even a photograph of such a monster rat been offered to back up these claims. Yet rats—an estimated eight million rats—do live beneath the streets of New York; which is a vast paradise for them. They are particularly fond of the insulation around Con Edison's high-voltage cable under Park and Fifth Avenues, finding the insulation so appetizing that they often eat it right down to the bare cable and are electrocuted, the tunnels reeking of fried rats for days after. As for huge rats, however, one Con Ed worker says, "I never met a guy yet seen a small rat. The rats they see are all big as cats—as dogs, even." As a point of information, the biggest rat yet weighed was a little under 3½ pounds; house cats average eleven pounds. Rats do bloat considerably when dead, an exterminator reports, but not that much.

## A Selection of American and British Expressions Employing the Rat*

*A rat's head.* A fool. British.
*Bull rat.* Police informer.
*Desert Rats.* Nickname of British soldiers in North African campaign, World War II, who were first so called by Mussolini.
*Company rat.* Company informer.
*Dock rat.* A bum on the waterfront.
*Fight like a cornered rat.* To fight furiously.
*Grease rat.* A dredge deckhand.
*Looks like a drowned rat.* Repulsive looking, disheveled.
*Pack rat.* A prospector; a petty thief (obsolete); a porter (little used).
*Rat.* A despised, distrusted, selfish, unethical person, often "dirty rat."
*Rat.* A doctor in training. Obsolete.
*Rat.* A loose woman. Obsolete.
*Rat.* A new college student.
*Rat.* An informer or squealer.
*Rat.* A nonunion printer. Obsolete.
*Rat.* A pad, shaped like a rat, used in certain women's hairstyles.
*Rat.* A pirate. British. Obsolete.

---

*As this collection shows, *rat* used by itself, as a suffix, or in slang phrases, almost invariably has a deprecatory connotation, meaning "unworthy," dishonest," "sneaky," etc.

*Rat.* A politician who deserts his party. British. Obsolete.
*Rat.* A railroad train, short for "rattler."
*Rat.* Nickname of a disliked Indiana religious sect in 1824.
*Rat cheese.* Inexpensive yellow cheese.
*Rat-face.* Someone sly and underhanded.
*Ratfink.* An informer, or any contemptible person.
*Ratpack.* Originally a teenage gang, then a gang of any kind.
*Rat race.* A way of life or job, etc., in which action or activity is more important than goals, or where "keeping up with the Joneses" is more important than satisfaction, like a lab rat on a treadmill.
*Rat Rime.* Doggerel. British. Obsolete.
*Rats!* An expression of annoyance or disgust.
*Rats.* A star (backwards spelling). British.
*Rats in the garret* (or *loft,* or *upper storey*). Eccentric or mad. British.
*Ratshit.* The worst.
*Ratso.* Derogatory nickname.
*Rat's tail.* A wit. British.
*Ratter.* An informer.
*Ratty.* Shabby.
*The Ratcatcher.* A nickname of Winston Churchill in World War I because he said of the German fleet: "If they do not come out to fight, they will be dug out like rats from a hole."
*The Rat's Hole.* Charing Cross Underground (subway) in London. British.
*To give one green rats.* To malign, slander, or backbite. British.
*To give one the rats.* To berate someone. Obsolete.
*To have rats.* To have the D.T.s, to be crazy, to be eccentric. British.
*To Rat-hole.* To hoard food.
*To rat on* (someone). To inform or squeal.
*To rat out.* To withdraw dishonorably.
*To smell a rat.* To suspect something bad.

# VIII

# BROTHER RATS
## A Rogues' Gallery of Rat Relatives

The Chinese hunters who came into the Gobi Desert after tarbagan weren't interested in local customs or taboos. They were poor and hungry, fleeing a famine farther west. They laughed when villagers warned them not to eat the lumpy fat under the rodent's shoulders because every tarbagan hunter was reincarnated as a tarbagan and these were the remains of the dead hunters. Neither did they pay much attention to prohibitions against eating tarbagans that appeared ill or against eating inadequately cooked tarbagan meat. But such folk wisdom had for centuries been useful to the Mongolians: sick tarbagans were often dying of bubonic plague; in which case the lymph nodes under their shoulders usually swarmed with plague bacilli. All the Chinese hunters of these tarbagans contracted bubonic plague, carried plague back to their villages, and started an epidemic that took sixty thousand lives in 1910 alone.

The tarbagan, a large marmot weighing up to nine pounds, is only one of scores of rodents that are bubonic plague carriers. Many other rat relatives are potential carriers—there are hundreds of species and subspecies. Indeed, judging by their many kinds, as well as their stupendous numbers, rats and their close relatives are already more the masters of the earth than man can ever hope to be. Distinct rodent species, a multitude of brother rats, are found in the New and Old Worlds, on every continent; in tropic, arctic, and temperate climates. To cover them all here would be impossible, but we can offer a representative sample from all corners of the world. Since rodent genetic engineering seems to have come into its own recently (see chapter 5) one can imagine what a monster might be created by selecting the best qualities of these brother rats: An aggressive destructive rat as large as the 174-pound capybara, with the intelligence and industry of a beaver, the breeding power of the bandicoot, the ferocity of the cotton rat, the speed and leaping ability of the kangaroo rat, the burrowing talents of the mole rat, the armor and longevity of the porcupine, the opposable thumb of the marmoset mouse. . . . Unlikely, of course. But the mind boggles, even H. G. Wells's mind would boggle. Why such a rough beast might even stockpile weapons like one other slightly rattish species does!

All rats are, in fact, stockpilers, or at least hoarders and collectors. The most talented in this respect is the eastern pack rat (*Neotoma floridana*), another plague carrier, which lives at virtually every altitude over a range from northern Canada to the Florida Keys. No other rat carries thievery to such extremes, and its name has become a synonym for a kleptomaniac junk-picking miser. The capacious nest of the pack rat is usually a large globe-shaped "castle" made of sticks, leaves, and grasses up to six feet high, resembling a small beaver lodge, but the nest can also be built in hollow logs, caves, abandoned cabins, in openings of trees, and even under cactus plants in the desert. In its museum or treasure-house nest are found every bright shiny thing one can imagine as well as some not-at-all shiny objects, including (and all these stolen objects have been documented) keys, eyeglasses, coins, currency, belt buckles, pens and pencils, lipsticks, watches, tinfoil, china, rags, socks, cartridge cases, bleached bones and skulls, sets of false teeth, and even mousetraps. The pack rat even has been known to steal coins from the pockets of sleeping campers. A legend persists that it will leave something in return for what it takes, but less romantic observers theorize that the little creature probably is passing by with something in its mouth when it comes upon a brighter trinket and drops whatever it is carrying to claim the glittering prize some strange compulsion commands it to possess. Nevertheless, the legend persists, which accounts for the pack rat's being called the trade rat.

True to its common name, the nocturnal pack rat lives in large packs or colonies, and when one individual is alarmed by an enemy such as an owl or coyote, it warns the others by stamping its hind feet on the ground and rapidly vibrating the tip of its tail, the noise carrying over a considerable distance. This rodent is about as large as a brown Norway rat but more resembles a squirrel with its soft grey fur, white underbelly, and large expressive eyes, which have flat lenses like a human's and must move in the socket (unlike the spherical-lensed eyes of most rodents, which focus an image equally well from any direction). One of the twenty-eight or so species, *Neotoma cinerea,* has a bushy tail like a squirrel and is thus called the bushy-tailed wood rat.

The pack rat will shriek when an enemy captures it and is known to make low chirps during the mating season. Mating occurs once or twice during the year and the male travels as much as a mile a night in search of a partner. After considerable trouble with the temperamental female, who is as apt as not to split his ear or bite his tail badly before submitting to his advances, the male remains with his partner longer than most male rodents do with their mates, sometimes staying in the nest some thirty-five to forty-eight days after the young are born.

Pack rats are excellent climbers and have a kind of stubby "thumb" on their front feet, as does the gray squirrel. The white-throated pack or wood rat (*Neotoma albigula*) is perfectly able to climb cacti without hurting itself and carries spiny needles from the cacti back to its nest to construct a prickly door at the entrance that discourages enemies from entering. In its nest are stored all its glittering treasures and food, including nuts, berries, seeds, fruits, leaves, insects, and other food it steals from any of man's fields or campgrounds nearby. Sometimes this intelligent creature actually dries out fresh greens in the sun before stocking them in one of its many castle storerooms.

Follow desert tracks that the naturalist Ernest Thompson Seton described as "a curious, delicate, lacelike fabric of polka dots and interwoven sinuous lines" and you may encounter another fascinating member of the rat family, the kangaroo rat (*Dipodomys*), of which there are fully five genera and about twenty species. Found in North American deserts and semideserts anywhere from Oregon and Colorado to Vera Cruz, the foot-long creature is shaped like a miniature kangaroo. It has proportionately the same short forelegs and long powerful hind limbs as its namesake, enabling it to jump as far as eight feet horizontally in a single leap. Its long strong tail helps it to maintain its balance in flight and even turn completely around in the midst of a leap. The tufted tail, longer than its body, and which it often uses as a third leg when standing, enables the kangaroo rat to catapult into the air and land on an object as small as a grasshopper three yards away. Leaping is its means of transportation no matter how long the journey, and the kangaroo rat even fights its battles in the air. A fight between two kangaroo rats literally takes place in midair, the combatants slashing at each other with their sharp hind claws. No quarter is given and the aerial combat ends only when one of the silent kangaroo rats is dead.

The little kangaroo rat lives alone except in the mating season, and, as its big round eyes indicate, is a nocturnal animal. Popularly styled "a galloping ghost of the desert," it lives in deep burrows with mazes of branching tunnels that provide a hiding place from coyotes, owls, and other enemies, and that also protect it from the harsh daytime desert sun. The burrows, closed off with dirt to discourage intruders, lead to storerooms where the kangaroo rat has been known to hoard more than twenty-five gallons of seeds. The little creature can carry thousands of such seeds in its mouth, which contains fur-lined pouches that can stretch to an extraordinary size. The kangaroo rat eats well and regularly, but it never drinks a drop of water or any other liquid in its two-year life span; it gets all the moisture it needs from the succulent prairie plants on which it nibbles. Its bigger cousin, the giant kangaroo rat, is likely to be put on the endangered species list soon.

Glamour star of the rat set is the beautiful black-and-white maned rat (*Lophiomyinae*) of Africa. When it is irritated, however, the maned rat raises the hair on its back, making it look like a porcupine and frightening its enemies. With its long ears, soft fur and blunt nose, *Leporillus,* native to Australia, is aptly named the rabbit rat. About the size of a baby rabbit, *Leporillus* builds nests out of piles of brushwood, nests so strong that even Australia's wild dog, the dingo, cannot break in. Another oddity is the spiny rat. Because of a weak link near the base of its tail, the South American spiny rat (*Proechimys*) often loses its tail and is sometimes called the tailless rat. This "weakness" serves the spiny rat well when an enemy grabs it by the tail; the rat can scurry off leaving only that tidbit in the mouth of its attacker. Much more dangerous is the North African gundi (*Ctenodactylidae*), a short-tailed guinea-pig-like rodent that became familiar to Westerners during World War II, when it proved to be the host for the rare disease toxoplasmosis, which was transmitted to soldiers, leaving a rash and affecting the lungs like pneumonia. So little is known about its ancestry that it is called "the forgotten rodent."

A rat one could call "Red" is the African rufous-nosed rat (*Oenomys hypoxanthus*), which has a reddish nose and tail base and whose upper body is brownish-red. A good climber, it builds its ball-shaped nest between tree branches or high stalks of grass. An African rival is the nocturnal giant-pouched rat (*Cricetomys*), so named because of its capacious cheek pouches, which have been known to hold as many as 275 Bridelia seeds the size of coffee beans. A thief and hoarder like the pack rat, this animal is valued as a food all over Africa.

The South American capybara (*Hydrochoerus*) takes the prize as the largest known rodent, weighing up to 174 pounds, standing over two feet tall and measuring 4½ feet long. Though as large as a small pig, the gentle vegetarian, also called the carpincho or water hog, would rather hide than fight and makes its home in tall grasses along river banks. Unfortunately it has many cunning enemies that it cannot always escape despite its swimming abilities. Its predators include the cougar, the jaguar, and man, all of whom hunt the giant rodent for food.

Among the largest of New World rats is the rice rat (*Oryzomys*), over 180 species of which inhabit the southeastern United States, Mexico, Central America, and South America. The rice rat is the common rodent of tropical America. A slender animal up to 12½ inches long, it varies in color from gray to russet, with short coarse fur, a long scaly tail, and often whitish feet. Ranging along the eastern seaboard from New Jersey to Florida and then west through all of the

Gulf states to eastern Texas, it is a serious enough pest, but not as destructive as the native cotton rat or any of the rats imported from Asia. Good swimmers, that actually prefer being in the water, rice rats often make their nests near the shore and they feed on seeds, fruit, fish, and invertebrates. Prolific, the female giving birth to up to nine litters a year, they are active both at night and in the daytime. Sometimes black rats make war against and destroy the weaker rice rat populations. The James Island rice rat (*Oryzomys swarthi*) of the Galapagos is believed to be the rarest rodent in the world—only four living specimens have been collected, and these in 1906. "The rice rat's name is something of a misnomer as it prefers green shoots and other foods to fish," wrote naturalist Irenaus Eibl-Eibesfeldt. "Every week it consumes twice its weight in food. Like the cotton rat it has no compunctions about eating one of its kind if starving."

Many years ago a scientist wrote of a plague of cotton rats in South America: "Everything that was not made of iron, stone or glass, whether it be furniture, clothing, hats, boots, or paper, everything carried the mark of these rodents' gnawing teeth. These rats even gnaw horn and wooden knife handles. They may ruin leather, cloth, or linen, destroy books and paper, and even test the strength and sharpness of their teeth on cherry pits. We do not know why they gnaw and bite through pewter vessels, lamps, lead balls and small shot. However, this is not all; cows have had their hooves gnawed even as they stood in their stalls, these rats have literally gnawed on fattened pigs, and they have even wounded sleeping people." Clearly a most undesirable character, the native New World cotton rat (*Sigmodon*), of which there are fortunately but three known species, ranging from the southern United States to Venezuela and Peru. Called the cotton rat because it often dines on cotton seeds, dropping the balls of cotton along its runways, it lives in low meadows and salt marshes, high up in the mountains, and at elevations between those extremes. This medium-sized rat is a major pest, devouring many cultivated crops, including alfalfa, sweet potatoes, sugarcane, and also eating the eggs of many birds, as well as young chicks. When there is no food the cotton rats quickly resort to cannibalism, killing and eating their weaker fellows. Hungry females sometimes go so far as to kill and gobble up their mates after the sexual act. Cotton rats are so numerous because they can reproduce all year long; the female, in fact, typically mates again a few hours after the birth of a litter and has a large litter of four or more every twenty-seven days of the year. Often millions of these sparsely haired, short-tailed, short-eared, buff-and-black creatures devour man's crops when conditions are favorable for breeding, five hundred of them to an acre is not at all an unusual number. The cotton rat destroys plants because it can't climb up to get the fruit; it digs the roots out of the earth or gnaws through the stem. Help against them comes from their natural enemies, snakes, foxes, weasels, and even alligators, as well as the epidemics that sometimes wipe out hordes of them.

Gourmet that it is, preferring tender young bamboo shoots to any other food, the bamboo rat (*Rhizomyidae*), is likewise a gourmet dish to many Asian people, who regularly hunt it for food. Six species of bamboo rat are native to southeastern Asia and eastern Africa. These rodents are thickset, up to fourteen inches long, and have prominent orange incisors. They are, of course, common to bamboo regions, and live in burrows that they dig with their teeth and feet. Bamboo rats spend a lot of time above the surface, feeding on grasses, seeds, and fruit as well as bamboo shoots. They are similar in looks and habits to the American pocket gopher. In Africa, natives hunt this "pig rat" with dogs, which drive the rodents out into the open where hunters club or spear them.

*Golunda,* the coffee rat, has greatly affected the worldwide price of coffee by destroying entire coffee plantations in Ceylon and India, among other places. Less destructive of crops, but more dangerous to humans is the multimammate rat or mouse. So named and called "the great mother of rodents" because of the many (12–24) nipples on the female, more than any other mammal except the pig, the multimammate rat (*Mastomys*) is a prolific animal which can give birth every four weeks and bear up to nineteen pups in a litter. At home in the steppes and savannas of Sub-Saharan Africa, where they live in colonies, the one-to-two-foot-long multimammates are able to live in the wild as well as near or in human dwellings. Typically spending the daytime in their burrows, they are active at night searching out grass, seeds, tubers, green shoots, and insects to eat. Luckily, considering their fecundity, they have many enemies, including owls, various cats, and snakes. These rats spread bubonic plague in parts of Africa, sometimes carrying the plague bacillus themselves, but usually merely hosting fleas that carry it. This is all the more unfortunate because they move into human settlements from the wild and then move from one human settlement to another. Despite their ferocity they are trapped and employed as laboratory animals for use in research on the plague, schistosomiasis, and cancer, from which they also often suffer.

There are seven known species of giant-tailed rats (*Uromys*), all of them with tails almost as long as their foot-long bodies. At home in the mountain forests of New Guinea and New Britain they sometimes live in trees and feed upon leaves, according to the natives, who hunt them for food while they sleep in the palms. The scales on the tails of all of these species are arranged in a mosaic pattern in row after row, very much like the species called the mosaic-tailed rat (*Melomys*). The mosaic-tailed rat boasts teeth so powerful that it can cut through a coconut shell to get at the meat inside.

The most regal of rodents is the Chinchilla (*Chinchilla laniger*), a native of the Andes often bred on farms today for its valuable fur. Chinchilla is the most valuable of all furs and chinchilla coats have sold for as much as $100,000, about 100 of the ten-inch-long creatures being needed to make a coat. The pale-blue or silver-gray animal has fur so fine that individual strands aren't visible to the

It takes 100 or more of these chinchillas to make a fur coat. Rat coats, hardly as valuable, used to be made from 300 rats' skins. *(Richard Lydekker, The New Natural History)*

Though it looks like a mouse, the little shrew isn't even a rodent. *(J. G. Wood, Illustrated Natural History)*

The agouti *(Buffon, Plates of Natural History)*

Common chipmunks can be plague carriers. *(Richard Lydekker, The New Natural History)*

The common dormouse, named for its sleeping habits *(Mayer's Lexicon)*

The gerbil, a prime plague carrier today *(Richard Lydekker, The New Natural History)*

Norwegian lemmings do not purposely commit mass suicide as is commonly believed, but many drown nevertheless. *(Richard Lydekker, The New Natural History)*

A musk cavy *(Century Dictionary)*

A water rat *(Buffon, Plates of Natural History)*

The Siberian pica; this rodent is little known in the West. *(Richard Lydekker, The New Natural History)*

The common kangaroo rat, at home in the American deserts. *(Richard Lydekker, The New Natural History)*

The European porcupine *(Mayer's Lexicon)*

Prairie dogs have spread plague in recent times. *(J. G. Wood, Animal Creations)*

A beaver rat *(Century Dictionary)*

(above) Naturalist Olaus Magnus depicted Norwegian lemmings falling from the sky to be devoured by ermines, in his history of northern nations, *Historia de gentibus septentrionalibus*, 1555.

ɘft) Colonies of marmots have caused ɒidemics of bubonic plague in the Gobi esert. *(Richard Lydekker, The New ɑtural History)*

Banner-tailed kangaroo rat *(U.S. Bureau of Sports Fisheries and Wildlife)*

Merrian's kangaroo rat of the western United States *(U.S. Bureau of Sports Fisheries and Wildlife)*

A herd of South American capybaras, sometimes called grunting carpinchos, the world's largest rodent at up to 174 pounds. *(J. G. Wood, Animated Creations)*

naked eye. Resembling a squirrel with a rabbitlike head, the chinchilla is slow to breed and produces only one or two litters a year.

Water rats (*Hydromyinae*), comprising several species, are the best adapted to aquatic life of all rodents, especially Monckton's water rat (*Crossomys monck-toni*), named for biologist Oldfield Monckton, with its watertight fur. Like the beaver, water rats live on the edges of rivers, bogs, lakes, and ponds. They should not be confused with Norway rats, which are sometimes called water rats. Good swimmers, they are essentially meateaters and hunt shellfish, crab, fish, frogs, and even water birds. The European water vole (*Arvicola terrestris*) is erroneously called the water rat, but is not really a rat at all. Actually a large brown field mouse, it is an accomplished swimmer and diver, feeding on vegetable matter, small fish, and crabs. The water vole has the unusual habit of sunning itself on platforms it builds of sticks and grasses among the waterside reeds. It carries a form of rodent plague, has been known to attack humans, and is a pest in gardens.

Many legends have attached themselves to the mole rat (*Spalax*). Russian peasants, for instance, believe that if you catch a Mediterranean mole rat, let it bite you, and shake it to death with your bare hands, you will be able to cure any illness by laying your hands on the afflicted person. A Libyan superstition has it that anyone who disturbs a mole rat will become blind. Mole rats are not closely related to the mole, and some species cause more damage to crops than moles do, by eating their roots. Their eyes are small and functionless and they are among the most developed rodent burrowers, using the head and powerful incisors as their basic digging tools. The fascinating mole rat or tuco tuco (*Ctenomys*) of South America is a large, ferocious creature resembling a pocket gopher; it actually emerges from the ground to attack sheep, grabbing them by the nose, when the sheep trample over its burrow. The African mole rate or bles mole is almost completely naked, with only a few hairs on its body. It is a socially organized creature whose members are divided into castes, all the diggers of tunnels being smaller rats, for example. The queen rat is the only female who breeds and she resides in a central chamber with her two or three mates and a number of large drone rats, who guard her. The other females can't breed because the queen releases an inhibiting substance in her urine, which the drones cover themselves with and spread throughout the tunnel system. Young male African mole rats actually beg feces from the drones because drone feces contain a substance that enables the young rats to digest cellulose plant material.

In recent times the lesser bandicoot rat has migrated from the fields to the villages and cities of India, where it has caused so much damage that it is now noted as one of the world's four most destructive rats, along with the Norway rat, the black rat, and the Polynesian rat. *Bandicota bengalensis,* native to southern Asia, is the most prolific of rats, the females often bearing a litter a month, with seven pups in each litter. Strong and robust, bandicoots are omnivorous, feeding on everything from green matter, poultry, and shellfish, to garbage. Known as "pig rat" in the Teluga language of India, they are burrowers, possessing strong claws and broad feet for digging in the earth. These nocturnal creatures often live separately underground and rob gardens of root vegetables. Good swimmers, they cause much damage in the rice paddies of India and Ceylon but, as mentioned, have proved even more dangerous in the large cities where they have migrated recently; there they are carriers of bubonic plague and other diseases. It is little compensation that they are hunted for food (see chapter 13), or that Indians often dig up the stolen corn that they store in underground "granaries" and use it for food.

Thirty centuries ago the little Polynesian rat (*Rattus exulans*) journeyed with Polynesian voyagers from Southeast Asia, possibly brought along as a living food supply for the adventurers. As noted, this species, responsible for recent plague epidemics in Vietnam, has since become one of the four most fearsome rats in the world; it now ranges from Southeast Asia all the way to New Zealand and Hawaii. It has the bad habit of taking one bite out of a sugarcane stalk, ruining it, and going on to the next and is thus as responsible for high sugar prices as any middleman.

According to American Indian legend, the muskrat lives in the marsh, neither land nor water, because it couldn't make up its mind if it wanted tò be a land or water creature when the sun god Nanabojou asked its wishes. This rodent (*Ondatra zibethica*) is trapped so extensively, some twelve million caught annually, that it provides North America's cheapest fur. Going by the aliases Hudson Seal, River Mink, Water Mink, River Sable, Bisam Mink, Electric Seal, Near Seal, Loutrine, and Hudsonia, among others, it is a prolific creature, the female often bearing thirty young a year. Muskrats are essentially water rats and feed on aquatic vegetation. They live in large mounds with underwater entrances built of rushes and reeds, doing considerable damage to earthern dams, dikes, and canals with their burrowing habits, in both America and in Europe (they were introduced to the Old World from Alaska in 1905). This beaver look-alike gets its common name from the odorous musk secreted by the male's inguinal glands. These male scent glands are used in perfume manufacture, while female muskrat meat, said to taste like good poultry, is sold in the United States under the trade name "marsh rabbit" or its Indian name *Musquash* to avoid any connection with rat meat.

The gerbil, a popular pet rodent today, is the principal wild rodent plague carrier in South Africa. Two of the most dangerous plague-carrying rodents in America are the squirrel and the marmot. As mentioned, the California ground squirrel was the first wild rodent to be infected with plague in the United States and probably passed the disease along to other animals, including various squirrel species. A large number of eastern fox squirrels were found to be infected in Denver, Colorado, in 1968, the first ever among city-dwelling squirrels (see chapter 3).

The marmot, native to North America, may have hosted plague thousands of years before the San Francisco outbreak, causing America's first bubonic plague, though most experts believe that the black ship rat from Asia was the culprit. Nevertheless, many marmots are definitely plague infected today and, unfortunately, are so tame in some parts of the West that they almost eat out of human hands. The hoary marmot was called "the whistler" by French-Canadian trappers because it often perches on high rocks to detect approaching danger and warns other marmots with a shrill piercing whistle.

Of the many unlikely animals closely related to the rat, such as the plague-carrying American prairie dog, and the beaver, the porcupine is among the most unusual. This spiny rodent is of the same family as the guinea pig. The porcupine, well known in antiquity, was the subject of an ode by the Roman poet Claudius; its sharp quills (and a mature porcupine has 30,000 of them) were once used as arrow tips and made into magic bundles and fetishes. Porklike porcupine meat, especially that of the crested porcupine, the largest species, is still eaten in Italy, Tunisia, Lebanon, Asia Minor, and South Africa, and fishermen still use porcupine bristles for making floats. The porcupine does throw its quills at enemies, though involuntarily. When cornered the animals erect their quills, shaking their bodies, and a quill or two can come loose with force. But the "terrible porcupine" of Shakespeare does not shoot hundreds of quills at dogs and hunters, as Aristotle and scores of cartoonists have had it from early times. Many a zookeeper has been shot by a loose quill and one farmer reported that a trapped porcupine had thrown several quills with such force that they became embedded in tree branches high above it. A porcupine quill tends to travel through the human body and can take thirty hours to pass from one side of the leg to the other. But despite its menacing barbed quills, large wild cats and trained dogs can overcome the porcupine. Porcupines are noted for their excellent sense of hearing and are said to be able to "recognize the characteristic sounds of edible materials falling to the ground," including the sound of an acorn falling from a tree several meters away. They live longer than any rodent; a crested porcupine in India holds the record of twenty-two years. There has been much speculation and joking about how the animals mate ("Carefully!"), but in reality the female simply flattens her tail against her back and the male mounts her with his forefeet. Newborn porcupines have soft bristles for only ten days and cannot be handled after that without great care. Nevertheless, more than a few people raise the creatures and report that they make faithful, intelligent pets that are easily housebroken.

Books have been written about the beaver (*Castor fiber*), the largest and most ingenious of North American rodents. Valuable beaver pelts, used largely in felt hats, were a great incitement to the exploration of the New World, and in the early nineteenth century hundreds of thousands of beavers were killed for their fur. Growing to a length of four feet and a weight of over eighty pounds, beavers are peaceful family animals devoted to their kittens, and live in large towns or colonies. Their engineering skills in building dams and lodges are well-known and they have been observed building 1,000-foot-long canals to divert a stream into their pond or to reach a food source (usually the inner bark of trees). Beavers have "more engineering skill than the entire Army Corps of Engineers," claimed one admirer in Civil War times.

The South American guinea pig (*Caira cabaya*) isn't a pig, doesn't look like a pig, and doesn't hail from Guinea in West Africa. Native to Brazil, it is a rodent that takes its first name from the fact that it was brought to Europe on the Guinea-men slave ships that sailed from Guinea to South America to deliver their human cargo and filled their holds with whatever cargo was available for the trip home. How the creature was mistaken for a pig is anybody's guess. Our slang expression "guinea pig" for anyone on whom something is tested, obviously derives from the widespread use of the guinea pig in scientific experiments. Easily raised, the guinea pig is reared for food by the Peruvian Indians. It is a popular pet and has saved many human lives through its use in the laboratory. The old story that rats won't enter a guinea pig cage is untrue; they not only will enter, they will eat the guinea pig's young.

Hamsters are today even more popular than guinea pigs as pets. Of the several hamster species common to Europe and Asia the best known is the golden hamster, a gentle and docile rodent that is much valued for laboratory research work. The golden hamster (*Mesocricetus auratus*), with its deep golden color, has the shortest gestation period of any nonmarsupial mammal, females giving birth in an incredible sixteen days. Just how widely this hamster has been studied by scientists is attested to by the three thousand articles about it cited in the *Bibliography of the Golden Hamster* by R. Kittel (1966). The millions of golden hamsters being bred today all descend from one hamster male and three females found by the scholar I. Aharoni in 1930. Professor Aharoni came across a brief reference in ancient Hebrew writings to a special kind of "Syrian mouse" occurring "in Chaleb which was brought into Assyria and into the land of the Hittites." Determining that Chaleb is today Aleppo in northern Syria, the Professor traveled there from his home in what was then called Palestine and found the "Syrian mice" that have since become known to millions as golden hamsters. The larger common hamster (*Cricetus cricetus*) of Europe is not so well known worldwide and has a much nastier disposition. This hibernating hamster lives in a very complex arrangement of underground burrows where it stores its food most systematically, separate foods in separate rooms. From four to eighteen young are borne by the females in each litter, but the mother has only four nipples and the extra young are bitten to death and eaten by the mother and by the remaining four sucklings. A menace to crops, and a plague carrier in South America, the hamster must be controlled by man. But it does kill field mice and other pests, its fur is used by furriers, and many people are fond of its meat. It is also one of those rodents whose stored food supplies are dug up by man; one writer has noted that "unemployed men would bring a rucksack of grain up out of the earth" from hamster burrows, and that even today people in Magdeburg, East Germany, obtain feed for their chickens from hamster stores.

Shrews, so often erroneously confused with rats and mice that they should be included here, are not rodents, but small mouselike mammals so ferocious that they attack and kill animals several times their size. The glands of a short-tailed shrew contain enough poison to kill 200 mice. A shrew's high metabolic rate forces it to eat almost continuously and it often consumes more than its own weight in a day. Shrews have become proverbial for their ferociousness and short temper. It is not as well known that they can literally walk on water thanks to their large hairy feet and light weight, or that the pygmy Etruscan shrew, with a heart that beats one thousand times per minute, is the world's smallest mammal, weighing about one tenth of an ounce. The little elephant shrew of East Africa, (*Elephantulus brachyrhyrcaus*), so named for its long snout, always places its feet in the same place in its extensive runs, following a far-reaching system of tracks that it initially made. Each foot invariably lands in the same footprint.

It would be hard to find a more interesting rodent to end with than the prolific lemming, which is among the most fascinating and mysterious of animals. One legend has it that lemmings commit mass suicide by drowning themselves (see chapter 2) and many people in Arctic regions believe that legions of the destructive little creatures fly through space from distant stars and fall to earth in heavy showers, to take the place of these "suicides." In fact, the Eskimo word for lemmings is *kay-loong-meu-tuk,* or "mice from the sky." Explorer-naturalist Sally Carrighar reported several lemming shower stories in *Wild Voice of the North* (1959). As with all rodent showers (see chapter 4) none was scientifically verified, but the lemmings could have been carried long distances by strong winds. "Several serious-minded Eskimos told me of seeing the lemmings come down, 'falling in bigger and bigger circles that turned the same way as the sun,' or clockwise," Carrighar wrote. "Eskimos who had not seen the lemmings descend, all could describe lemming tracks that 'start where the lemmings landed without any footprints going out to that place.' The late Reggie Joule, an Eskimo bush pilot, said that the familiar spurs of lemming tracks often are found on the roofs of the Eskimo cabins at Point Hope, where he grew up, 'and there weren't ever any tracks outside those cabins.'"

## Rodent Relatives

*Order Rodentia*

**Suborder Sciuromorpha** (squirrel-like). Seven families, including: squirrels, prairie dogs, gophers, chipmunks, beavers.
**Heteromyidae** (pocket mice). **Perognathus** (pocket mouse) **Dipodomys** (kangaroo rat).
**Suborder Hystricomorpha** (porcupine-like). Sixteen families, including: porcupines, guinea pigs, chinchillas.
**Suborder Myomorpha** (mouselike). Ten families, including:
**Family Cricetidae** (ancient mice; "the squeaking ones"), including:
- **Cricetinae** (New World mice and rats)
- **Peromyscus** (white-footed mouse)
- **Onychomys** (grasshopper mouse)
- **Reithrodontomys** (American harvest mouse)
- **Sigmodon** (cotton rat)
- **Oryzomys** (rice rat)
- **Neotoma** (wood rat)
- **Cricetus** (hamster)
- **Microtinae** (lemmings and voles)
- **Microtus** (meadow and field mouse)
- **Clethrionomys** (red-backed mouse)
- **Ondatra** (muskrat)
- **Gerbillinae** (gerbils)

**Family Muridae** (modern mice, "the mousy ones"), including:
- **Murinae** (Old World mice)
- **Rattus** (house rat)
- **Rattus rattus** (black rat)
- **Rattus norvegicus** (brown rat)
- **Rattus exulans** (Polynesian rat)
- **Mus** (house mouse)
- **Micromys** (European harvest mouse)
- **Apodemus** (yellow-necked mouse & European long-tailed field and wood mice)

**Family Gliridae** (dormice), including:
- **Glis** (bushy-tailed dormouse)
- **Muscardinus** (hazel mouse)

**Family Zapodidae** (jumping mice), including:
- **Zapus** (jumping mouse)
- **Neozapus** (jumping mouse)

# IX

# OF MICE AND MEN
## A Mouse Is Miracle Enough

"As I had never seen a mice plague" an Australian rat historian wrote "and was skeptical of the reports of babies nibbled in their cradles and of cats too full to bother hunting I drove down to investigate for myself. There really was a plague. The verges and fields were riddled with holes and runways . . . and at night, if one went out with a torch [flashlight], the mice were hopping about under foot. I parked a caravan [mobile home] in the area and began writing the last chapter of this book. But mice have invaded the caravan. They are running over the table as I write, fighting at my feet for the Cornflakes, and last night their noise and their invasions of my sleeping bag made sleep impossible. . . . It seems a better way to end like this; to just close the typewriter and throw away the rest."

The beleaguered writer was describing a mouse plague near Port Wakefield, Australia in a book he wrote about rodents. He was not the first to discover that the common house mouse can be as dangerous to man as the rat, despite man's ambivalent and even benign attitude toward it. The little mouse can be an even greater danger than the fiercer rat because its smaller size enables it to hide and thrive where the rat cannot. The mouse carries the same infections as the rat, including bubonic plague, enteritis, tularemia, salmonella, and many other viral and fungal infections. It also spreads botulism, foot-and-mouth disease, swine erysipelas, choriomeningitis, and trichinosis. Weil's disease, or leptospiral jaundice, a wasting disease of the human liver that can vary in intensity from mild flu-like symptoms to a fatal form of jaundice, is a constant danger from mice. Mice tolerate the leptospira, a group of spiral micro-organisms, in their kidneys, where they cause no harm. When transmitted to man these leptospira become dangerous, and a drop of infected mouse urine containing them can be fatal on entering a small cut or abrasion on a person's hand.

In addition to the food they eat and ruin, house mice have done great damage to crops since classical times, when many "mouse plagues" were recorded. In 1789 Australian plagues of house mice that had migrated to the fields destroyed millions of tons of food on several occasions, and in 1917 thousands of *tons* of mouse carcasses (sixty thousand mice to the ton) were gathered in that country. Six million mice were killed in a single night in one Australian town, and a six-week campaign in another area netted six hundred tons of them. On two occasions, 1926 to 1927 and 1941 to 1942, over seventeen field mice per square yard, or eighty-two thousand mice to the acre, were counted in the central valley of California, the soil appearing as if it had been tilled by them in their efforts to find food. Some 14 million mice were killed in Monterey County, California, in the summer of 1968.

Early in 1982, a smaller plague of mice infested the newly restored California state capitol; while the restoration proceeded for six years, the mouse population exploded, and when the legislators returned to their desks they encountered a virtual parliament of mice. A legislative assistant found a mouse swimming in his coffee cup and at least ten mice were killed in the office of State Senator William Campbell, who captured one by stepping on its tail as it scampered past his desk.

Even American presidents aren't immune to mice troubles. President Jimmy Carter spotted two mice scampering across the carpet in his little hideaway White House office and notified the General Services Administration (GSA) who are the federal housekeepers. Before anything was done, however, one of the creatures died inside the walls near the Oval Office, bathing the area with the odor of dead mouse at a time when Latin American heads of state were due in Washington to sign the Panama Canal treaties. The other mouse continued to run free about the office and nothing was being done about it or the smell. In typical bureaucratic fashion, the GSA claimed that the mice were "outside mice" and not their domain—that they were wards of the Interior Department, which is in charge of the White House grounds. Naturally, the Interior Department disagreed and the White House still stank. "I can't even get a mouse out of my office," the President moaned. Mr. Carter finally had to line up two top officials from each agency in the reeking Oval Office to break the bureaucratic deadlock.

Members of the same genus as rats, mice are so closely related to their larger more feared first cousins that a common house mouse (*Mus musculus*) artificially inseminated with rat sperm will produce hybrids of the two animals. This is something scientists haven't been able to do with black and brown rats. The major differences between the rodents is that the mouse is less intelligent or certainly less suspicious than the rat, getting caught in baited traps any self-respecting rat would disdain. The rat is also much larger, typically several times heavier and often ten times as long. Generally speaking, rodents with a body and tail length of thirteen to fifteen centimeters and weight of eighteen to thirty grams are called mice while rodents measuring and weighing more than this are considered to be rats. Mice may indeed have evolved from rats at a time when it was more advantageous to be able to hide by squeezing into a smaller hole than to be large and aggressive.

True mice of the genus *Muridae* probably did not emerge until late in Miocene times. *Mus musculus,* the best known of mice, is found throughout the world today, but this little house mouse, like the rat, originally lived on the Asian steppes in southern Russia, Iran, and Turkestan. From these plains of central Asia, it spread through the Middle East to Europe and from there to the rest of the world. The house mouse reached America at roughly the same time as the Norway rat—not until about 1700, when it stowed away aboard trans-Atlantic ships and came ashore with their cargo.

One scientist has told how field mice moved into his camp on an uninhabited island after they had been living outside for more than eighty years. An extremely adaptable animal, the mouse can survive from the tropics (in the huts of natives) to Antarctica (in the huts of scientists living there), and in the worst of conditions outside and inside. House mice have been discovered living eighteen hundred feet down in coal mines, feeding on the scraps of miners' lunches. They flourish in totally dark cold-storage meat lockers where the temperature is at most 17 degrees Fahrenheit (-80 Centigrade). These last mice eat the meat stored in the lockers, grow thicker fur and become heavier. They often make their nests in the meat carcasses, and—hardly frigid in sex—produce more offspring every year than mice living in houses.

We take the species name *Mus musculus* for the common house mouse from the Romans, whose word for "little mouse" was *musculus.* Their word for mouse also gives us the word for muscle (and the French, Spanish, Italian, German, Danish, and Swedish words as well). Some whimsical Roman probably noticed that the working, rippling muscles of athletes resembled little mice appearing and disappearing in play, and so he named the muscle after the mouse. The Romans also named the marine mussel we eat after *musculus,* in this case because of its size and color.

About 130 species of house mice have been described, but all of the common house mice belong to the one *Mus musculus* species. These are the western house mouse (*Mus musculus domesticus*), native to western and northwestern Europe, west of the Elbe, which lives largely in the buildings of man; the northern house mouse (*Mus musculus musculus*), found east of the Elbe, which spends part of the year outdoors, weather permitting; and the eastern house mouse (*Mus musculus spicilegus*) which lives outdoors all year around, only occasionally boarding with man. Most house mice depend on human communities for their food supply. This was dramatically shown on the tiny island of St. Kilda off Scotland's west coast, where a special population of house mice thrived for many years. When St. Kilda's human inhabitants were evacuated from the island in 1930, the entire mouse population died out.

House mice are said to be able to live up to six years, but the oldest house mouse known, nicknamed "Hercules," lived five years eleven months, from January 1971 to December 26, 1976, in Purley, Surrey, England. Usually mice die in their first winter, though many live eighteen months or more. The mouse makes up somewhat for its smaller size and lower I.Q. by being even more prolific than the rat, starting to breed at about six weeks old. Male house mice are quite obsessive in courting estrous females, but some of their relatives are no less than amazing. Stuart's marsupial mouse (not a true mouse) stays mounted—in place on a sometimes sleeping female—for as long as twelve hours, while the gerbil-like Shaw's jird had been observed making 224 mountings in the course of an hour (see chapter 4).

House mice can breed in complete darkness or in temperatures well below freezing. It appears that the male mice produce an odor that stimulates females to undergo the hormonal changes that ready them for sexual intercourse. Mice will copulate any time, any place, the male mounting the female from the rear for a few seconds, but nature has provided them with a rather unique type of birth control when the house mouse population becomes too dense for the food and/or space available. Group restlessness at these times produces hormonal changes in the females, particularly adolescent females, making them infertile. Specifically, the vagina of such a female will not open to a male mouse, the uterus becomes very thin, and the ovaries become inactive.

Pregnant female mice carry their young nineteen to twenty-one days, and the female can become pregnant again two days later. A litter contains five to seven young on the average, though litters of thirteen and more have been recorded.

In recent laboratory tests, mouse embryos were grown successfully for ten days under completely artificial conditions in laboratory dishes; they grew to the stage where the limb buds were clearly visible and the minute bubblelike formations that would develop into eyes and ears could be seen clearly under the microscope.

Litter size generally depends on where the house mouse is living. Inside houses the average litter contains about 5.5 young; in cold-storage lockers, almost 6.7; in flour warehouses, 7.8; and in corn ricks, 10.2. Female house mice bear five to six litters a year, usually in the normal mating period from March to October. The young are born naked, deaf, and blind, weighing about one gram, and measuring half an inch. One could fit, curled up, inside a wedding ring. The eyes at birth are like dots and the ears, minute marks. The babies' wrinkled bright pink skin begins to show fuzzy hair after five days and is completely covered with hair at ten days, their eyes and ears opening at from thirteen to sixteen days after birth. At this time they are fully equipped mice, complete with their efficient incisor teeth. They begin fending for themselves a bit by the end of the third week, occasionally leaving the mother, although they continue to nurse from the mother until the end of the fourth week. One of the dangers they face are male mice, even the father, who often will kill young babies, especially in captivity, sometimes neatly biting the heads off their little pink bodies. When under great stress the mother too will kill her young, eating some or all of the litter.

The well-built nests made by house mice are usually constructed of soft, dry vegetable matter such as hay, straw, burlap, rags, cardboard, and even insulating material that the mother chews up to make it more pliable and which she painstakingly carries piece by piece to the holes or cracks in the building where her nest will be situated. Where space is limited, housemice share their nests, and in these communal nests as many as fifty babies are mixed indiscriminately, mothers with weaned offspring often helping to nurse other nestlings. In any case, the mice usually live in groups, each group or super family consisting of a breeding pair and their children and their children's children. Each family occupies a common territory and marks it off from other packs by delineating its borders with urine scent marks that warn off intruders, and which account for the "mousy" smell common to mice-infested places. Unrelated male mice frequently fight until one emerges as the leader of the group and life becomes peaceful except when strangers intrude on a group's territory. In his delightful book *Mice All Over,* which records his controlled study of captive wild mice for the British during World War II, biologist, Dr. Peter C. Crowcraft describes such an encounter:

> The owner of the territory rushed at the intruder and knocked him over or bit him on the rump or tail. The intruder fled across the border and the attacker stopped running after him. On breaking off the chase the attacker often worked off his anger in the hunched-up "pussyfooting" gait. . . . Safely within his own territory, the other mouse turned to face the direction of his former attacker and rattled his tail violently. . . . If an attacker pressed his attack into the territory of the retreating male the fleeing mouse would turn and rear up on its hind legs. This was the signal for the attacker, now off his home ground, to retreat at once. . . . Obviously in this kind of situation rearing up . . . signalled to the attacker . . . "Here I stand and here I will give battle." Rearing up is common in prolonged battles between mice and indicates a high state of aggressiveness, just as raising one's fists is indication of a higher state of aggressiveness in humans than is clenching the teeth.

Though murder seems to be as rare among mice as among other wild animals, Crowcraft did observe two cases where intruding mice were killed with bites through the back of the skull. He observed that all mice will fight. Tame laboratory mice thrown in with wild mice will be peaceful at first but then their heritage prevails, and they soon produce the same type of society with a tyrant heading it. Even infant mice will attack a stranger that enters their nest, although they will run away from him outside the nest.

Once a mouse wins a fight, that initial advantage becomes psychological, and he tends to maintain it, "nothing succeeds like success" apparently as true among mice as among men. Any male mouse who wasn't victorious when introduced into a territory by Crowcraft had a terrible time of it, surviving "only because of his companions in misery [other defeated mice], all of whom dissipated some of the dominant's aggressiveness." There was in fact, a whole underclass of mice:

> Some males which had been badly defeated and bitten, and were no longer able even to put up resistance, resigned from the mouse race and became social parasites or tramps. . . . They became so cowed that they no longer displayed aggression towards one another, but huddled together for warmth. . . . They did not sleep side by side, but in layers, each one trying to get underneath the others. . . . These tramps escaped the attention of the dominant mice. They had found, by painful trial and error, the vacuums of activity which occur in even the most crowded situations; those odd corners which were not included in the habitual paths of the patrolling mice. Only in an artificial environment, with plenty of food near at hand, and no predators, could these pariahs survive. But given those conditions they hung on tenaciously to life, scruffy, dirty, and extremely smelly.

A bit more optimistically, Crowcraft does point out that a mouse can begin a new life in a virgin territory and succeed, in time recovering its "mousehood," much like human colonists do in a new land. And mice aren't totally tyrannical and selfish. Escape holes, urine areas, and defecation areas, as well as nesting and eating sites, are all shared collectively by the mouse group. Mice in a group also practice mutual grooming, which is an important part of the mouse's daily routine, one mouse cleaning another's back, neck, and all other areas the individual isn't able to reach itself.

Adult house mice generally have brownish gray fur with paler underparts and a tail thicker and scalier than other mice. Males and females are almost identical. Though mice have relatively large eyes in proportion to their bodies, sight is the least important of their senses, helpful primarily in detecting movement at close quarters. Their sense of hearing is far more acute and is attuned to picking out high-frequency sounds. Human ears cannot hear some of the sounds mice make; unfortunately for the mice, however, the house cat can hear every last one. What mouse sounds humans do hear sometimes resemble music. There have been reports from all over the world telling of "singing mice" with voices weak but birdlike in quality, their song "a series of pleasing musical chirps and twitters." Experts investigating such claims have usually found abnormal conditions in the mouse's nose and throat and attribute the "singing" to bronchial and asthmatic conditions. Mice also have been reported to appear at concerts and stand stock still listening to the music. This "musicality" has been explained as their indifference to tones from the lower register, which sound muffled to them and don't trigger a flight response.

The mouse's keen senses of smell and taste aid it greatly in its nighttime prowling for food. Its long muscle whiskers (*vibrissae*) are extremely sensitive and help guide it through narrow spaces in dark unfamiliar places. Mice are not strictly nocturnal animals, feeding mainly at dusk and dawn, but eating sporadically during the day as well. Their activity area may not be more than fifty square feet or so for their entire lives, but they can do anything they have to do within this area.

Mice can run, jump, and even swim, though they don't go into the water willingly. Extremely good climbers, balanced by their long slender tails, they can scamper up and down almost any vertical surface. A mouse can find its way back to its nest after searching for food by smelling the trail laid down by minute glands in the soles of its feet, which secret an oily substance. But this is only the case when the mouse has the time to do so. When in flight to its shelter, a mouse relies upon what is called its kinesthetic sense, its subconscious memory of sequences of muscular movements.

If food is scarce in its domain, the house mouse will travel distances, which are great relative to its size, up to a quarter of a mile a week, to obtain food from warehouses, granaries, and other sources. Interestingly, when spring comes, even in suburban areas, it will move outside into the fields until the warm weather ends in fall. Equipped with the sharp chisel-like incisors rats and all rodents possess, the house mouse is able to gnaw through most materials with ease. It eats many of the same foods as people do, including grains, protein foods such as meat and cheese, and fruits and vegetables, but has been known to consume almost anything, including soap, glue, plaster, and chalk. While living outdoors, it feeds on grass and weed seeds, plant stems and roots, and even worms and grubs. In its original native habitat upon the Asian steppes, the mouse was almost strictly a seed feeder, searching for tiny ripe grass seeds scattered on the earth. Feeding was a matter of eating a little at a time at frequent intervals, which became part of its adaptation for its original way of life. The mouse, after thousands of years, still retains its ancestors' feeding behavior; in one enclosed area experiment where grain was spread densely all over the ground, an individual mouse was seen to feed in at least 893 different places in three weeks. Because of this feeding habit the mouse is more difficult to eradicate with poisons than the rat. Mice tend to feed so sporadically in time and in so many places that they are unlikely to eat a lethal dose of poisoned feed at one location. Instead, a mouse will more often take a sublethal dose of the poison, and develop symptoms that may be agonizing but not fatal, which warn it away from the food.

A full-grown mouse eats little more than four grams of food a day, but it contaminates far more in the process with its droppings and urine, rivaling the rat in the amount of food it ruins and dangerous diseases it spreads (see chapter 2). A great number of natural enemies limit its destructive behavior, including the house cat, the owl, and the weasel, which is small enough to enter mouse burrows to kill, and neatly turns the mouse's skin inside out when it eats the creature. Even the spider can snare mice. In one recorded case, a black widow spider bound a mouse in its web, administered a lethal bite, and hoisted the mouse into the air, where it remained suspended until the spider was ready to dine.

Despite all the damage they do, mice are indeed timid, as our sayings "meek as a mouse," "mouselike," and "mousy" indicate, their infinitesimal hearts, weighing but 1.15 grams and beating 620 to 780 times a minute. They are easily shocked. More than one mouse has dropped dead from the sudden loud noise of a snap-trap mousetrap after avoiding capture.

Man, in his attempt to eradicate the house mouse, has been building better mousetraps for centuries (see chapter 6), but none has succeeded and the house mouse is still with us in vast numbers. So are its various country cousins who live outdoors all year round, whose numbers are truly staggering. It would be impossible to treat the hundreds of mouse species in this short space, but we should take a look at some of the more important and interesting ones from most of the ten families of the suborder *Myomorpha* (mouselike creatures), including the New York mice (*Cricetinae*) and the Old World mice (*Murinae*). Incidentally, the New World mice are considered to be the "ancient" mice in evolutionary terms, while the Old World mice common to Europe are the "modern" mice.

First country cousin to the house mouse is the field or meadow mouse (*Microtus*), often called the vole in Europe and sometimes called the wood mouse. There are about 238 types of field mice ranging throughout the Northern Hemisphere, including North America and Europe, many of them plague carriers. From earliest times, the field mouse has done great damage to crops in mouse plagues, forming an intricate network with their burrows under fields. Their population fluctuates on a four-year cycle and at the peak of each cycle the ground is literally covered with kestrels, buzzards, owls, gulls, crows, magpies, shrikes, herons, storks, weasels, polecats, foxes and many other predators feeding on nothing except these mice. Yet their numbers decrease mainly because of sheer nervous and physical exhaustion, the result of their overcrowding.

A Hungarian scientist kept a pair of common field mice, and in ten months the pair and its offspring of three generations produced 2,557 animals, reproducing more quickly than butterflies, which lay 200 to 300 eggs a year. Female field mice are generally still nursing a litter when they become pregnant. Among the most abundant of all rodents, the field mouse is the fastest breeder of all rodents, the meadow vole species of Britain (*Microtus agrestis*) producing as many as 17 litters a year with 13 young in each litter (121 offspring), females becoming pregnant while they are still sucklings 25 days old.

The field mouse lives underground in burrows and its legions have often destroyed an entire year's crop in an area. The creature is so voracious that it has been known to invade beehives, kill the bees, and eat the honey. This house mouse look-alike can swim as well as it can climb and dig, as this charming selection from *Riverby* (1894) by the great American naturalist John Burroughs shows:

> I met a little mouse in my travels the other day that interested me. He was on his travels also, and we met in the middle of a mountain lake. I was casting my fly there when I saw just sketched or etched upon the glassy surface a delicate V-shaped figure, the point of this V was being slowly pushed toward the opposite shore. I drew near in my boat, and beheld a little mouse swimming vigorously. . . . His little legs appeared like swiftly revolving wheels beneath him. As I came near he dived under the water to escape me, but came up again like a cork and just as quickly. It was laughable to see him repeatedly duck beneath the surface and pop back again in a twinkling. He could not keep under water more than a second or two. Presently I reached him my oar, when he ran up it and into the palm of my hand, where he sat for some time and arranged his fur and warmed himself. He did not show the slightest fear. It was probably the first time he had ever shaken hands with a human being. He was what we call a meadow mouse, but he had doubtless lived all his life in the woods and was strangely unsophisticated. How his little round eyes did shine, and how he sniffed me to find out if I was more dangerous than I appeared to his sight. After a while I put him down in the bottom of the boat and resumed my fishing. But it was not long before he became very restless and evidently wanted to go about his business. He would climb up to the edge of the boat and peer down into the water. Finally he could brook the delay no longer and plunged boldly overboard, but he had either changed his mind or lost his reckoning, for he started back in the direction he had come, and the last I saw of him he was a mere speck vanishing in the shadows near the other shore.

About the only thing beneficial to man the field mouse seems to have achieved is the result of its attacks on hyacinth bulbs in Holland. Dutch growers found that when the bulbs were gnawed through to the heart by mice they broke up into innumerable bulblets that grew into new bulbs, which taught them to increase their supply of valuable hyacinths by slashing the bulbs to the heart and cutting cross sections in them. Aside from this the field mouse appears to have no redeeming features. Luckily for us his life span is a mere one to one and a half years and he is found delicious by as many predators as any creature on earth. Field mice are eaten by cats, lynxes, weasels, skunks, badgers, wolves, dogs, foxes, coyotes, bears, birds from hawks to seagulls, all kinds of snakes, even by snapping turtles, frogs, and fish. But even this would not be enough: they might still overrun the earth if it weren't for the fact that their population becomes hopelessly overcrowded and dramatically declines after reaching a peak.

Another great pest, the American harvest mouse (*Reithrodontomys*), looks even more like the house mouse than the meadow mouse in size and appearance. A great acrobat, it builds a weatherproof spherical nest high up in tall swaying grasses. The harvest mouse's mating call is a high-pitched buglelike song. Its close relative the Old World harvest or dwarf mouse (*Micromys minutus*) is much smaller, weighing as little as 0.15 of an ounce, an even smaller form being found in Russia. These harvest mice search for food and eat for three hours, then sleep for three hours, repeating the process all day long. When the wind blows they can be seen riding the high grasses, swaying back and forth, close together yet somehow never hitting each other.

Most beneficial to man is the grasshopper mouse or scorpion mouse (*Onychomys leucogasten*) of western America, for although it is a plague carrier this brownish red seven-inch-long rodent feeds mostly on insects—grasshoppers, scorpions, beetles, lizards, cicadas—and even cannibalizes relatives such as the pocket mouse, when it can subdue them. Only a few other mouse species, notably the Congo forest mouse (*Deomys ferrugineus*), are mainly insect eaters. The grasshopper mouse is very fierce, not at all "mouselike," and barks rather than squeaks when on the trail of prey. A night hunter, it scents out its kill with its nose close to the ground like a dog's and even wags its tail when it picks up a scent! It will attack anything, leaping up at its quarry's throat and killing with its outsize incisors. Grasshopper mice can eat one-half their body weight in a single day. But these fierce, voracious creatures are kept as pets; some people let them loose at night to eat roaches and other household pests, which ought to have earned them the title of the exterminator or pied piper mouse.

Beneath the ground, mole mice (*Myospalacini*), which are as big as hamsters and similar to moles in their habits, destroy many crops in their range from Russia to China by gnawing off the roots of plants. Living a large part of its life underground, the mole mouse uses the large claw on the third finger of its forefoot as a digging tool and can dig a large seventy centimeter tunnel in less than twelve minutes. This rodent snarls when trapped. It is hunted in China as a pest, but Chinese farmers use it as a weather predictor, claiming there will be good weather as long as mole mouse tunnels are open, bad weather when they are closed by the mole mice. Another molelike mouse pest is the pine mouse

(*Pitymys*) common to the pine forests of the eastern United States. The pine mouse rapidly reproduces and has ruined many a crop of potatoes in a short time; it even eats the root systems of fruit trees, often destroying an orchard. Still another burrower is the Australian native mouse (*Leggadina hermannasburgensis*), of which there are some two hundred related species in southern Asia. These yellowish brown or brown-gray mice are unusual in that they collect and store pebbles in their burrows. It is thought that the pebbles collect all the water the mice need to survive in their arid desert environment.

It has recently been discovered that a climbing arboreal species of mouse, native to Costa Rica, pollinates the epiphyte Blakea, an air plant with clusters of bell-shaped flowers that grows in the tops of trees. The climbing mouse inserts its snout and tongue inside the flowers and sucks out the nectar, getting a dusting of pollen on its nose that it delivers to the next flower.

Climbing mice are found in many areas of the world. The African climbing mouse (*Dendromus insignes*) is one of many climbing types of the African rain forest, though it lives high up on Mount Kilimanjaro, too. This black-striped rodent can use its prehensile tail imperfectly for support and gripping, often climbing trees and bushes searching for seeds. The long-tailed climbing mouse (*Vandeleuria oleracea*), with a tail longer than its chestnut brown body, also favors trees; one was reported to make its nest high up in a strong spiderweb still used by spiders. The tree-climbing marmoset mouse (*Hapalomys*) of Indochina actually has the hands of a minute monkey with an opposable thumb for taking hold of tree seeds and fruit. Females of the red tree mouse of northwestern U.S. forests live in nests high up in the trees, while the males live in burrows under the earth. Male and female consent to meet only in the mating season.

Mice that can make prodigious leaps in the air include the Australian hopping mice (*Notomys*), ten species with powerful hind legs that hop like a kangaroo when alarmed and can easily leap three feet up, hopping sideways as well when they want to. The woodland hopping mouse (*Napaezapus*) seems to jump up and down doing an erratic dance during the mating season. Birch mice (*Sicistenae*) are closely related to jumping mice, but are more noted for their eight-month hibernation period, the longest of any mammal. Champion jumper of all the jumping mice is the kangaroo mouse (*Zapus*), a hibernator that weighs only about an ounce, has a tail twice as long as its little body, and can jump twelve feet in a single leap.

Best known of the jumping mice are the jerboas (*Dipodidae*) of African, Middle Eastern, and Central Asian prairies and deserts. The jerboa, another plague carrier, hibernates for five months and seems to get by without drinking any water all year. It often lines its nest with its own breast or belly hair, though it uses camel hair, too. The male jerboa delights in whacking the female on the snout during its mating dance, in which it continually circles her, challenging her to assume the mating position. Big-eared jerboas of Chinese Turkestan have

Field mouse *(Century Dictionary)*

Rice field mouse *(Century Dictionary)*

Harvest mice near their nest *(Louis Figuer, Mammalia)*

Short-tailed field mouse and long-tailed field mouse *(H. E. Bates, Through the Woods)*

The common meadow mouse *(Century Dictionary)*

The red-backed meadow mouse *(Century Dictionary)*

The common house mouse *(Mus musculus) (Joe E. Brooks, U.S. Public Health Service)*

An old woodcut entitled "Cat and Mice"

Mice assembling around their king are the basis of many tales, such as this one illustrated by Ernest Griset for his book *Little Folks,* 1882.

This sixteenth-century woodcut depicts an allegory of the Trojan War, a fantastic battle between mice and frogs.

one-and-a-half-inch ears on a three-inch body. All the many jerboas are kangaroo-shaped with a tuft of hair on the tail that acts as a steering mechanism in their long leaps through the air. They land on padded feet to cushion their fall and have been seen walking off slowly on their two hind legs.

Smallest of the mouse family is probably the three-inch pygmy mouse (*Baiomys taylori*) of Africa, which nevertheless has a voracious appetite and eats anything. Only the shrew, among mammals, is smaller than this pygmy. Among the fattest mice is the African fat mouse (*Steatomys*), composed of twenty-five subspecies, a rotund creature that can store fat in its body and is the only African mammal that hibernates. The African fat-tailed mouse (*Pachyuromys*) is a gerbil-like sparsely haired animal of a shimmering flesh-pink color that stores fat in its tail for use when food is scarce.

Common but unusual in its habits is the white-footed mouse (*Peromyscus*) of North America, where 178 kinds, including the plague-carrying deer mouse, are found. For unknown reasons this "traveler" mouse moves its home from place to place very capriciously. It is sometimes called the vesper mouse because it is very active in the evening when vesper church services are held. A musician who sometimes indulges in prolonged humming, it is also said to drum with its feet on hollow objects such as seeds. Owls are dangerous enemies of the white-footed mouse. An owl will swallow a mouse and digest all its flesh, but cannot digest the bones and fur. This is regurgitated in the form of a round pellet, consisting of every little bone in the mouse's body, covered with fur. Such pellets are often found in the fields.

"The mice without mothers," spiny mice (*Acomys*), sport stiff flattened spines on their dark brown backs that are thought to give them protection against cactus thorns or insect bites in their desert home. Unlike the young of other mice, their babies are not born blind or helpless. At three days old they are shakily exploring their surroundings and do not need a true nest. In the extended sense of the word they have no single mother. Midwives or "aunts" help a mother spiny mouse give birth right in the midst of the pack, beginning to lick a baby clean when it is only half out of the mother. After one to three days the young become the common property of all the nursing mothers in the pack.

Many mouse species illustrate the principle of protective coloration. The banana mouse (*Dendromus*) of Africa, for example, is yellowish with a dark stripe down its back, which enables it to hide in folds of banana leaves or among the fruit. The striped grass or zebra mouse of Africa (*Lemniscomys*) has a striped black tail that makes it easier to hide in the grass. An extreme example is the pocket mouse of the White Sands National Monument in New Mexico. *White* pocket mice there have white fur in order to blend in among the white desert sands. In black lava nearby are *black* pocket mice which blend in with the rock. Finally, there are red earth hills in the vicinity in which live *red* pocket mice.

Among the most amazing members of the mouse family is the dormouse, which looks more like a little squirrel than a mouse and lives as long as six years. Noticing that this chipmunk-sized rodent hibernated during the winter, someone far back in history dubbed it the dormouse from either the French *dormir,* meaning to sleep, or the Old English word *dorm,* meaning to doze, which also gives us the word dormitory. In fact, traditionally the dormouse, often called the Spanish rat or tree rat, has been regarded as one of the "seven sleepers" of the animal world along with the ground squirrel, marmot, hedgehog, badger, bat, and the bear.

A hibernating dormouse, coiled up with its forefeet tucked under its chin, its hind feet clenched in front of its face and its tail curled over its face on to its back, can be rolled across the table like a wheel and still won't wake up. The savage garden dormouse will immediately devour any dormouse that begins its winter sleep before the others, even its own mother.

The fat or edible dormouse (*Glis glis*) has been widely eaten in Europe since Roman times, when the Romans actually raised it for food (see chapter 13). It used to be thought, however, that the dormouse was poisonous. The naturalist Topsell explained this old superstition in his *Fourfooted Beasts*:

> If the viper find their [the dormouse's] nest, because she cannot eat all the young ones at one time, at the first she fillith with one or two, and putteth out the eyes of the residue, and afterwards bringeth their meat and nourisheth them, being blind, until the time that the stomach serveth her to eat them everyone. But if it happen that in the meantime any man chances to light upon the viper-nourished blind dormice and to kill and eat them, they poison themselves through the venom which the vipers hath left in them. Dormice are bigger in quantity than a squirrel. It is a biting and angry beast.

Lewis Carroll's fantasy of Alice trying to keep the drowsy dormouse awake at the Mad Hatter's tea party is better based on fact than Topsell's viper-blinded dormice. But there are so-called blind dormice, a Southeast Asian species that lives in logs and would seldom have to use its eyes. Other unusual dormice include the rock dormouse of Africa, which has an oddly flat body and skull that enables it to crawl into narrow rock crevices, and the garden dormouse, which leads its young in caravan fashion everywhere, the young always tailgating it. Most unusual is the spiny dormouse (*Platacanthomys lasiurus*) of the Indian Malabar Coast. The spiny dormouse has nasty quills on its back and natives of the area say that cats will not eat it—nature's catproof mouse!

The common house mouse itself is the direct ancestor of all the so-called "fancy mice" types that can often be bought in pet stores, including albino white lab mice, introduced from Japan in the early twentieth century; spotted mice; curly-tailed mice; long-haired mice; wavy-haired mice; and even hairless mice, which grow white fuzz after birth, then shed it, and are completely hairless like their parents within seven to ten days. The house mouse is also the direct descendant of waltzing or dancing mice. A hereditary middle-ear deficiency disturbs this mouse's sense of balance and it can't move in a straight line, but always weaves about in loops and circles. Walking backwards, however, it can walk an almost straight path.

Aside from its value as a laboratory animal since the beginning of the twentieth century (see chapter 11), there is little to be said in defense of the mouse—unless it be that when mice devour table crumbs they help to reduce the numbers of other vermin such as cockroaches! Yet, strangely enough, world literature, the best reflection of man's feelings, celebrates the mouse more than it castigates him, and certainly regards the mouse with nothing like the hatred and contempt with which we regard the rat. Mice are usually depicted as tiny, weak, timid and shy creatures, not vicious and destructive. They are underdogs who symbolize the plight of all of us at one time or another in varying degrees. As Dr. Crowcraft observed, "There is something terribly familiar about the awful situation of the mouse in the world." The Scottish poet Robert Burns best expressed this in his famous poem "To a Mouse, on Turning Her Up in Her Nest with the Plough, November, 1785." The poem is reproduced here with translations of Scottish or archaic words in brackets so that all can understand it without a specialized dictionary:

Wee, sleekit [sleek], cowrin, tim'rous beastie,
O, what a panic's in thy breastie!
Thou need na start awa sae hasty,
    Wi' bickering brattle! [flittering flight]
I wad be laith [loathe] to rin an' chase thee,
    Wi' murd'ring pattle! [plough staff]

I'm truly sorry man's dominion,
Has broken nature's social union,
An' justifies that ill opinion
    Which makes thee startle
At me, thy poor, earth-born companion,
    An' fellow mortal!

I doubt na, whiles [at times], but thou may thieve;
What then? poor beastie, thou maun [must] live!
A daimen icker [a corn ear now and then] in a thrave [shock]
    's a sma' request;
I'll get a blessin' wi' the lave, [rest]
    An' never miss't!

Thy wee bit housie, too, in ruin!
Its silly wa's [weak walls] the win's are strewin!
And naething, now, to big [build] a new ane,
    O' foggage [aftergrass] green!
An' bleak December's winds ensuin
    Baith [both] snell [biting] an' keen!

Thou saw the fields laid bare an' waste,
An' weary winter comin' fast,
An' cozie here, beneath the blast,
    Thou thought to dwell—
Till, crash! the cruel coulter [plow share] past
    Out thro' thy cell.

That wee bit heap o' leaves an' stibble,
Has cost thee mony a weary nibble!
Now thou's turn'd out, for a' thy trouble,
    But [without] house or hald, [dwelling place]
To thole [bear] the winter's sleety dribble,
    An' cranreuch [hoarfrost] cauld! [cold]

But, Mousie, thou art no thy lane, [alone]
In proving foresight may be vain;
The best-laid schemes o' mice an' men
    Gang aft agley, [go off the right line, awry]
An' lea'e us naught but grief an' pain,
    For promis'd joy.

Still thou art blest compared wi' me;
The present only toucheth thee:
But, och! I backward cast my e'e,
    On prospects drear!
An' forward, tho' I canna see,
    I guess an' fear!

That is the best thing ever written about mouse and man, but there has been much more written before and after Burns' best-laid plans. In the way of poetry, Walt Whitman wrote in "Song of Myself" that "a mouse is miracle enough to stagger sextillion of infidels," while W. H. Auden wrote a celebrated poem "Talking to Mice," and Alan Dugan wrote the touching "Funeral Oration for a Mouse" in which he wrote of bravery and hunger in a "Living diagram of fear" who "full of life himself . . . brought diseases like a gift."

Chaucer, Shakespeare, Swift, and Pope, are only a few of the English poets who have fitted the five-inch mouse into poetic feet. "Anonymouse" hath been even more prolific about the wee creature. There is the crooked man whose crooked cat caught a crooked mouse; the pussy cat that frightened a little mouse under the chair; and of course:

Hickory, dickory dock
The mouse ran up the clock,
The clock struck one
The mouse ran down;
Hickory, dickory dock.

The last, telling as it does of the mouse's penchant for late hours and its timidity, is rivaled only by the jingle "Three Blind Mice" as a popular rhyme:

Three blind mice, three blind mice,
See how they run, see how they run!
They all ran after the farmer's wife,
Who cut off their tails with a carving knife,
Did you ever see such a sight in your life
As three blind mice?

Several historians have linked "Three Blind Mice" with three men whom England's Queen Mary I put to death during her reign, the farmer's wife being "Bloody Mary" herself. But the song hardly lives on because of this all-but-forgotten incident in history.

Mice have been the subject of myth and fable since prehistory. Egyptians of old believed that mice were begotten from the mud of the Nile when the heat of the sun baked it in summer. The Greek philosopher Aristotle claimed that mice were born from dirt in ships and houses. For centuries in Greece, Sicily, and Asia Minor the god Apollo was honored as the ratcatcher or mousekeeper Smintheus, and believed to punish errant farmers with "mouse plagues." The mouse was indeed more highly esteemed in early Greece than it ever had been or has been since, people keeping white mice in their temples to help them seek the advice of the gods and regarding it as a good omen when the white mice reproduced abundantly. "Little mouse" was a pet name frequently appearing in Greek drama, and the mouse became symbolic of tenderness and sensuality.

Mice were thought by the Romans and other ancient peoples to cure ailments as varied as constipation, goiter, snakebite, epilepsy, baldness, and cataracts. Pliny the Elder found it "a pretty medicine" to cure headaches "by kissing only the little hairy muzzle of a mouse," and claimed that "If little infants wet their beds, a ready way to make them contain their water is to give them sodden mice to eat," a remedy that can't be found in any edition of Dr. Spock's. He also observed that "When a building is about to fall down, all the mice desert it," a precursor of the belief that rats desert a sinking ship. Pliny's fellow Roman, Plautus, made the mouse proverbial when he wrote, "Consider the little mouse, how sagacious an animal it is that never entrusts its life to one hole only," which is the forerunner of the proverb, "The mouse that hath but one hole is quickly taken," common to many languages.

Fully seven stories feature the mouse in *Aesop's Fables,* which were allegedly invented by a black slave belonging to Iadmon of Greece and were written down in the sixth century B.C. In "The Mice in Council," Aesop tells of several young mice who propose to bell a dangerous cat, until a wise old mouse asks which one is going to do the belling. "It is one thing to propose, another to execute," reads the moral. The story has since been told in many versions, sometimes with rats doing the belling.

Aesop's "The Cat and the Mice" tells of wise mice who aren't fooled by a treacherous cat who hangs herself from a peg in the closet pretending to be dead. "He who is once deceived is doubly cautious," is the moral.

In "The Lion and the Mouse" a mouse saves a lion, as he promised he would when the lion spared his life, demonstrating that "No act of kindness, no matter how small, is ever wasted."

"The Country Mouse and the Town Mouse" describes the temptations of a country mouse who leaves his simple safe home for the sumptuous but perilous life of his city cousin, proving that "A crust eaten in peace is better than a banquet partaken in anxiety."

The mouse in Aesop's "The Mouse and the Frog" learns his lesson when he lets a frog make a journey with him into the country and loses his life as a result. The mouse is also taught a lesson in "The Mouse and the Weasel" when he greedily eats too much in a big basket of corn and can't get out through the hole in the basket.

Aesop's "The Mountain in Labor" tells of people watching a mighty earthquake in a huge mountain, waiting "with bated breath" to see what will happen next, when a huge fissure appears and a tiny mouse pops out of the gap. The moral is "Magnificent promises often end in paltry performances."

During the Middle Ages it was believed that the soul exited at death through the mouth of a person in the form of a mouse. A red mouse indicated a pure soul at the time and a black mouse a soul blackened by pollution. Mice, like rats, were generally regarded as creatures of the devil in medieval Europe, and the devil and witches were believed to be able to change into mice at will. Mice were hated for the damage they did to crops at the time and in 1519 the farmers of Stelvio, Italy, brought mice to court charging that they had destroyed their crops. An attorney was appointed to defend the mice and he argued that they did some good by eating harmful insects. The judge, however, on hearing all the evidence sentenced every mouse in the area to immediate exile from Stelvio, the mice with young ones given an extra two weeks to leave. He even ordered the townspeople to keep their cats inside while the mice left and ordered small bridges built over streams to enable the mice to cross. History does not record if the verdict changed anything in Stelvio.

The famous medieval proverb "poor as a church mouse" means poor indeed, as church mice in centuries past had lean pickings, if they had any pickings at all. Churches then had no recreational facilities with well-stocked kitchens like their modern counterparts. They were used only for religious services and a mouse living in one would be hard put to find a crumb. The saying goes back to at least seventeenth century England, but was taken from similar German and French expressions that are much older.

According to a tradition preserved in the *Magdeburg Centuries,* in the tenth century Archbishop Hatto of Mainz, a noted statesman "proverbial for his perfidy," was devoured by mice. In 970, so the story goes, there was a great famine and Hatto, councillor of Otho the Great, decided that there would be more food for the rich if he got rid of the poor. Assembling all the poor in a great barn at Kaub he burned them alive, remarking, "They are like mice, only good to devour the corn." The legend claims that "God sent against Hatto a plague of mice," and to escape the mice he retreated to a town on the Rhine near Bingen. But the mice army swarmed upon his tower by the hundreds and thousands and "at last he was miserably devoured." A tower still stands near Bingen today and is called the Mouse Tower. In reality, however, it was built by Bishop Sigreied two hundred years after Hatto died, as a toll house for the collection of duties. The German word *Maut* means "toll," while the German mouse is *Maus,* the similarity of the words probably giving rise to the tradition of the tower. Similar tales are told of Bishop Widerolf of Strasburg, who is said to have been eaten by mice when he suppressed the convent of Seltyen in 997, and Bishop Adolf of Cologne in 1112.

On his supposed voyage from Ireland to discover America between 565 and 573, St. Brendan is said to have found an island where the mice were as big as cats Mice as big as horses figured in the fairytale "Cinderella," at least Cinderella's fairy godmother sent her to find six mice which she turned into beautiful horses.

In the old "A Frog Went A-Courtin'" the frog courts a Lady Mouse:

> There was a frog liv'd in a well,
> And a merry mouse in a mill.
>
> This frog he would a-wooing ride,
> And on a snail he got astride.
>
> He rode till he came to my Lady Mouse hall
> And there he did both knock and call.

Quoth he, "Miss Mouse, I'm come to thee,
To see if thou can fancy me."

Quoth she, "Answer I'll give you none,
Until my Uncle Rat come home."

Uncle Rat does give his permission, but the unlucky frog is swallowed by a duck before he can marry the pretty mouse.

Shakespeare mentions mice many times in his plays, including *King Lear,* in which "the fishermen that walk upon the beach appear [from a distance] like mice." The mouse is indeed proverbial in English, from the anonymouse "Quiet as a mouse," "It is a bold mouse that nestles in the cat's ear," and "When the cat's away the mice will play," to "Play cat and mouse with," "Watch him as a cat would watch a mouse," and Saki's "In baiting a mousetrap with cheese always leave room for the mouse."

Universally famous is Ralph Waldo Emerson on mousetraps: "Build a better mousetrap and the world will beat a path to your door." This, however, is found nowhere in Emerson's writings. Mrs. Sarah Yule claimed in a book published in 1889 that she had heard Emerson say it in a public address, his exact words: "If a man can write a better book, preach a better sermon, or make a better mousetrap than his neighbor, though he builds his home in the woods the world will make a beaten path to his door." In an 1855 selection from his *Journal* Emerson does write the following, which mentions no mousetrap and is really quite different from the commonly accepted proverb: "I trust a good deal to common fame, as we all must. If a man has good corn, or wood, or boards, or pigs, to sell, or can make better chairs or knives, crucibles or church organs, than anybody else, you will find a broad hard-beaten road to his house, though it be in the woods."

Mousetraps are mentioned wonderfully in Lewis Carroll's *Alice's Adventures in Wonderland* (1865): "They drew all manner of things—everything that begins with an M . . . , such as mouse-traps, and the moon, and memory and muchness. . . ." And American poet Amy Lowell described herself as "Hung all over with mouse-traps of metres . . ." As every child knows, the mouse also figures in Clement Moore's "The Night Before Christmas" (1823): "'Twas the night before Christmas, when all through the house / Not a creature was stirring, not even a mouse . . . ," the last image probably influenced by Shakespeare's "not a mouse stirring" in *Hamlet,* as every perfect wayward mind knows.

In *The Wizard of Oz* the Tin Woodsman saves the queen of all field mice from a wildcat and she repays him by assembling a great army of mice to drag the portly Cowardly Lion out of a deadly poppy field where he would have slept forever if he hadn't been moved to safety. Not as well known is the contemporary tale of Mouseland. Mouseland is a nation of intelligent mice just off the Great Barrier Reef in Australia that was discovered by shipwrecked Marvin Lampkin at the turn of the century. After being captured by the enterprising mice,

Lampkin found that they had been granted human intelligence because they released a god from bondage years ago and that their society was much like that of humankind. The mice held Lampkin prisoner because they realized that their nation would not survive once the outside world learned of them. But Marvin, made a citizen of the little country, managed to escape these Lilliputians during a Mouseland war and was rescued by a passing ship. . . . The whole story is an obvious imitation of Jonathan Swift's *Gulliver's Travels,* though one of the oddest tales anyone ever wrote. It can be read in full in Edward Earle Childs' *The Wonders of Mouseland* (1901).

There are probably hundreds, if not thousands, of stories and cartoons featuring mice today, with the mice always the hero, ranging from the inimitable Mickey Mouse (see chapter 10) and the all-powerful Mighty Mouse to Jerry of "Tom and Jerry" fame, the underdog or undermouse who always wins against great odds. In those last cartoons Jerry sometimes winds up sleeping alongside Tom Cat. This phenomenon actually has been reported from time to time. Syd Rodinovsky, professor of zoology at Millerville State College in Pennsylvania, actually did find a cat caring for a mouse that nestled against her like the rest of her litter of week-old kittens; what's more, he filmed the entire scene so there can be no doubt of it.

A good mouse tale to end with is the story of a mouse that could make one as rich as Mickey Mouse made Walt Disney. It is the little-known Scandinavian folktale of "The Tide Mouse," which is best recounted in the words of William A. Craige in his *Scandinavian Folk Lore* (1896):

> If a person wishes to get money that will never come to an end, one way is to produce a tide-mouse, which is got in this way. The person takes the hair of a chaste maiden, and out of it weaves a net with meshes small enough to catch a mouse. This net must be laid in a place where the person knows that there is treasure at the bottom of the sea, for the tide-mouse will only be found where there is silver or gold. The net need not lie more than one night, if the spot is rightly chosen, and the mouse will be found in it in the morning. The man then takes the mouse home with him and puts it wherever he wishes to keep it. Some say it should be kept in a wheat-bushel, others say in a small box; it must have wheat to eat and a maiden's hair to lie upon. Care must be taken not to let it escape, for it always wants to get into the sea. Next, some money must be stolen and laid in the hair beneath the mouse, and then it draws money out of the sea, to the same amount every day as the coin that was placed under it; but one coin must never be taken, otherwise it will bring no more. One who has such a mouse must be careful to dispose of it to another, or put it back into the sea, before it dies, otherwise he may suffer great harm. If the man dies, the mouse returns to the sea itself and causes great storms on sea and land; these are known as "mouse storms."

# A List of American and British Expressions Employing the Mouse*

*A mouse hunter.* A wencher. British.
*As drunk as a mouse.* Very drunk. British. Obsolete.
*As sure as a mouse tied with thread.* Very uncertain. British.
*As warm as a mouse in a churn.* Very snug. British.
*By the mouse-foot!* A mild oath. British.
*Mouse.* A bruise on or near the eye.
*Mouse.* Affectionate term for an attractive young woman, or sweetheart, wife. *Mouse.* A barrister. British.
*Mouse.* A harlot arrested for brawling. British.
*Mouse.* A mustache.
*Mouse.* A small military rocket.
*Mouse.* A small rat (q.v.) for a woman's hair.
*Mouse.* A timid informer, squealer.
*Mouse.* Be quiet!
*Mouse.* The penis. British. Obsolete.
*Mouser.* A detective. British. Obsolete.
*Mouser.* A mustache.
*Mouser.* The female pudenda. British. Obsolete.
*Mouse-piece.* In beef or mutton the part just above the knee joint. British. Obsolete.
*Mousetrap.* A submarine. British.
*Mousetrap.* The mouse. Obsolete.
*Mousetrap.* The female pudenda. British. Obsolete.
*Mousetrap.* A cheap nightclub or theater.
*Mousetrap.* A play to trap a would-be tackler by the offensive line in football.
*The parson's mousetrap.* Marriage. British.
*To mouse.* To hunt patiently and carefully.
*To mouse.* To fake an opponent out of position, to fool or mislead.
*To mouse.* To neck. Obsolete.
*To mouse over.* To sample, dip into.
*To speak like a mouse in a cheese.* To speak faintly or indistinctly. British. Obsolete.
*Trouser mouse.* The penis.

---

*Compare with "rat" expressions (pp. 152–53), noting how much less derogatory are these phrases.

# X

# AN AUTHORIZED* BIOGRAPHY OF THE REAL MICKEY MOUSE (*Via Firm Agreements with His Heirs)

Pablo Picasso kept a pet white mouse in a drawer in his studio that he sketched on occasion, but Walt Disney is far and away the artist most of us would associate with mice, Disney having created what is surely the most famous mouse (or rat) in history—Mickey Mouse. In over a half century on the screen this inspired rodent's hold on the public imagination throughout the world hasn't weakened any—children still revere him, adults collect him, people (with good reason) are still voting for him in general elections. He'd be middle-aged, fifty-five or more if alive, a star who ranks with Chaplin in stature and influence, among the greatest movie, stage, and television performers in our time. Yet no complete, impartial biography has been written of the most popular performer in all history, doubtless the most famous of all animals endowed with human characteristics. Certainly "Mouse, Mickey," as at least one scholarly index lists him, deserves a panegyric somewhere, so at the risk of being called misanthropic and anthropomorphic, following is the first historette of a mouse. Biographies have been written about far less appealing creatures.

To begin with, it's certain that El Ratón Miguelito was a *real* mouse like

Picasso's. For the sake of convenience we'll call our Mouse "him," though the original may just as well have been a miz as a Mister. Mickey Mouse (b.1923) came into this world somewhere within the musty, malodorous, cobwebbed walls of the garage that served as the late Walt Disney's Laugh-O-Gram studio in Kansas City, Missouri. The ur-material is elusive, but he appears to have come from a family of ten, though no accurate description of his ancestors is available. Yet to those cynics who question his reality, we submit abundant testimony that our hero (at first christened Mortimer) did exist. Disney's daughter once confirmed this in a long interview. "Several stories have been spread about Father's having a mouse who lived on his desk during his early days in Kansas City," she told reporters. "The thought back of this tale is that the mouse had given Father a special fondness for mice. 'Unlike most of the stories that have been printed,' Father told me, 'that one is true. . . . Mice gathered in my wastebasket when I worked late at night. I lifted them out and kept them in little cages on my desk. One of them was my particular friend. Then before I left Kansas City I carefully carried him out into a field and let him go.'"

Another Disney confidant advises that the artist let his mouse loose in an empty lot; exactly which house stands over it now in K.C. remains something of a mystery. Anyway: "Nine mice skittered off into the weeds, but the tenth stayed put. It was Mortimer, watching with bright eyes. Walt stamped his feet and shouted. The mouse took fright and ran. 'I walked away,' Walt would later recall, 'feeling such a cur.'"

Other sources go so far as to claim that Mortimer trespassed on his master's drawing board, "cleaning his whiskers with unconcern or hitching up his imaginary trousers;" that he plagued Disney's cartoonists to the extent of gnawing their pencils and erasers; that Disney often brought two lunches to the office, one for himself and the other for his pet—at a time when he more than once had to scrounge stale bread from restaurants for his own meals. There is even a tale that Disney forbade his employees to set traps for *any* marauding mice, keeping his favorite Mortimer in an inverted wire wastebasket during the day and letting him romp free at night. Here we have the familiar story of the neglected artist in his garret, the variation being that this neglected artist starved with the rodent who was to become the inspiration for his greatest creation. "Other people would leave lunch scraps in the wastebaskets," Disney later testified. "What *I* didn't eat I'd let the mice come around to eat. One (Mortimer) was bolder than the rest. There was a shelf above my drawing board and he wouldn't move off it."

There, then, are all the hard-core facts relating to the Mouse's real life before Disney left Kansas City for Hollywood with just four crumpled ten-dollar bills in his pocket. But we can speculate that Mortimer was about six months old when freed in that empty lot, and judging by the one-and-a-half-year lifespan of micekind, lived until August 1924, unless famine, feline or other bad fortune befell him. He could have had as few as nine brothers and sisters, but then again the common house mouse bears five to eight litters a year, so the family was doubtless larger, say fifty sibling rivals. As for his vital statistics, Mortimer probably measured a little smaller than the average *Mus musculus* at 2½-3 inches in length. His heart, however, though weighing only the usual, infinitesimal 1.15 grams, was decidedly larger in the way of soul, inspiring Disney to whatever greatness he achieved.

No one had ever created a Mouse anything like Disney's. The artist's fascination with rodents can be traced back to the days when he lived on a Missouri farm and used to draw pictures of mice on the barn walls with the five-cent box of crayons his Aunt Margaret had bought him. He was eight years old then and has been quoted as saying that at the time, "There was a man named Clifton Meek who used to drawn cute little mice, and I grew up with those drawings. . . ." Clifton Meek was the creator of America's first mouse cartoon, a four-panel strip called "Johnny Mouse" published in about 1912 by the Scripps-MacRae syndicate, but Meek's mouse was nothing like the maestro's.

The idea for the Mouse of Mice actually came five years after Disney freed Mortimer in that lot in Kansas City. To be exact, it came on the evening of March 16, 1928, aboard a train carrying the cartoonist from New York to Hollywood, when Disney dreamed of his all-but-forgotten pet. Call it what you like: fate, Lady Luck, serendipity, but while Disney dozed that night another world was in the making. The artist didn't know it, but he was gestating a mouse realer than real. Mortimer was actually put on paper for the first time on the next day, somewhere between Toluca, Illinois, and La Junta, Colorado. Disney at first drew him with ruffled hair like Charles Lindbergh's, for his first cartoon, *Plane Crazy,* was to parody that great adventurer, who had just flown the Atlantic, the plot revolving around a mouse who built a plane in his own backyard. But the familiar red velvet pants with red buttons, the black dots for eyes, the pear-shaped body, pencil limbs, cumbersome yellow clodhoppers, and three-fingered hands in white gloves—all were present from the beginning. So was a tail, which the rarest of rodents lost to an eraser in the future. Disney doodled and sketched all day. Suddenly, in the midst of his efforts that evening, he shouted to his wife Lillian: "I've got him—Mortimer Mouse!"

But wife Lillian contended—insisted—that Mortimer was a horrible name for a mouse. "Well then, how about Mickey?" Disney countered. "Mickey Mouse has a good friendly sound."

Lillian agreed, suggesting Minnie Mouse as a mate for Mickey in the process, and the Mouse was born. Prênom "Mickey" now, patronym still "Mouse." It was both a second life, a reincarnation for Mortimer probably long in his grave and a new birth for Disney, who even became the Mouse's squeaky voice when sound films were made by Mickey (later Jimmy MacDonald and then Wayne Allwine took over). "I fathered him when he was called Mortimer Mouse," the artist once said, "and he was my first born and the means by which I ultimately achieved all the other things I ever did—from Snow White to Disneyland."

Disney produced both *Plane Crazy* and *Gallopin' Gaucho* before he was able to sell Mickey Mouse in the black-and-white talkie *Steamboat Willie,* the world's first animated sound cartoon. But it wasn't all that easy. The first efforts with Mickey were silent cartoons. Disney himself did none of the drawings for these or any other Mouse opuses, delegating them to his assistant, artist Ubbe (Ub) Iwerks, who was with him almost from the start. The late Ub Iwerks always insisted, however, that Disney could easily have done the drawings if he hadn't been too busy. Mickey proved fairly simple to draw. His head and ears, for instance, were uncomplicated circles and his three white-gloved fingers and thumb were only depicted that way because, even while anthropomorphizing the mouse, four digits were easier and cheaper to draw than five. Ub Iwerks commandeered Disney's Hollywood garage and worked on *Plane Crazy* for six weeks, sometimes turning out seven hundred sketches a day until he finished, assisted only by a few young girls who traced the drawings over in india ink before they were filmed.

Mickey lived on film now, yet it remained a long journey to the hearts of millions of his fans-to-be. *Gallopin' Gaucho* (parodying the silver-screen feats of Douglas Fairbanks) came next, and even before the third and most famous Mickey film, *Steamboat Willie,* had been finished, Disney rushed to New York to exhibit the team's first two efforts. To his dismay, distributors showed no interest. New York theaters had all been equipped for sound with the success of Al Jolson's *Jazz Singer* in 1927. No one wanted a mute mouse movie and Disney returned to Hollywood.

The team decided to remake *Steamboat Willie* in sound, a project that seems easy now but which had never been done before. Furthermore, the cartoon conversion had to be done on an almost nonexistent budget. Could cartoon characters be synchronized with sounds recorded before the drawings were even completed? And what about sound, anyway? There wasn't money for an orchestra, nor even for actors to read dialogue. The technical synchronization trouble was solved after some hard thinking and the sound problem proved no problem at all. Disney, clamping his nostrils tight with his fingers, became Mickey's squeaky falsetto voice, a role he was to play for the next eighteen years. As for the music, this had to be supplied by Ub Iwerks on a washboard, Disney operating a sliding whistle, artist Wilfred Jackson playing a harmonica, and the whole crew clanging, blowing, and banging and whistling on whatever else was available. New Year's Eve noisemakers, cowbells, ocarinas, and a whole assortment of diverse instruments were brought into play. The musically talentless trio stuck with it until they completed the film, until one evening Disney saw the cartoon and kept repeating aloud, "This is it! We've got it!"

The New York distributors agreed. *Steamboat Willie,* first booked on Broadway at the Colony Theater, opened on September 19, 1928, and ran for a solid two weeks before it moved on to the Roxy as a featured attraction. Mickey, with his squeaky voice and jerky walk, played the captain of a Mississippi River steamboat who danced and tooted his boat whistle with gay abandon. "Mickey growls, whines, squeaks, and makes various sounds that add to the mirthful quality," *The New York Times* film critic wrote in applauding the cartoon. Other critics all over the world joined in praising Disney's creation. With *Steamboat Willie*'s appearance in 1928 the Mouse became an overnight sensation.

1928: A year of superlatives (a million shares a day changing hands on the stock market, the Dempsey-Tunney fight, Bobby Jones winning his sixth Open Championship, Shipwreck Kelly beating all flagpole-sitting records . . .) in which the most superlative of all species was the Mouse. That year Mickey began to win fame enough to satisfy the most ambitious *Homo sapiens.* His popularity knew no national boundaries; sojourners everywhere found home in his cartoons. In France he became known as Michel Souris; in Italy as Topolino; in Japan, Miki Kuchi; in Spain, Miguel Ratoncito; in Latin America, El Ratón Miguelito; in Sweden, Musse Pigg; in Germany, Michael Maus; and in Russia, Mikki Maus.

Much more was to come. By 1931, Mickey Mouse clubs in the U.S. had a membership of over one million. In 1935 the League of Nations gave Mickey Mouse an unprecedented vote of approval. Mussolini loved Topolino, and England's King George refused to go to the movies unless they starred Mickey. When Mickey skittered up on the podium to shake hands with the real conductor Leopold Stokowski in the film *Fantasia,* it was inevitably Stokowski whose reputation was enhanced, not the Mouse. There was only one exception: Hitler's propagandists called Mickey "the most miserable ideal ever revealed . . . mice are dirty."

With the advent of the Mickey Mouse Club there were tens of millions of Mouseketeers, a generation of young Americans and their parents eventually knowing the words to that often nerve-wracking theme song with the refrain, "Mickey Mouse, Mickey Mouse–M–I–C–K–E–Y, M–O–U–S–E" as well as they did the national anthem. New York's conservative Metropolitan Museum of Art hung an original portrait of Mickey Mouse. The Mouse appeared in twenty foreign newspapers, was enshrined in Madame Tussaud's wax museum; *Film Daily* estimated that over 100,000 people a day saw him on the screen. Mary Pickford, long America's sweetheart, announced that Mickey was *her* sweetheart, and Mohamet Zahin Kahn, potentate of Hyderabad, called him the leading American hero in India. At its peak the Mouse's fan mail approached ten thousand letters a day.

Mickey took the world by storm, "a small-town boy bashful over girls," in Disney's words, whose adventures were hardly namby-pamby (in one Rabelaisian episode he got his foot caught in a bedpan), but who was never to utter a stronger expression than "Oh, shucks." In 1935 King Features syndicate distributed the first Mickey Mouse comic strip and Mickey Mouse comic books won millions of devoted readers. Lionel Corporation discharged itself from bankruptcy by selling 253,000 Mickey Mouse toy handcars in one Christmas season. The Ingersol Watch Company sent its five millionth Mickey Mouse watch to Walt as a token of its appreciation; the watch became so popular that one Disney animator discovered it could break the language barrier in Outer Mongolia. The riches that the "Mouse that Turned to Gold" brought to Disney, galactic sums indeed, can be seen in a photograph *Life* published in 1953 showing three thousand of the then thirty-five hundred items which had been allowed to use Disney trademarks.

Mickey prompted Disney to create a host of fantabulous anthropomorphic cartoons and characters. "Father Goose," as he has been called, inspired creations that will never be forgotten. You could count on the knuckles of your fingers folk heroes as appealing as Disney's nanoid creations: Pluto, that most amiable of clumsy dogs; Donald Duck, his voice supplied by actor Clarence Nash; Goofy, played by Pinto Colvez, later Bozo the Clown for Capitol Records; Scrooge McDuck; Horace Horsecollar; José Carioca; Clarabelle Cow; and even a relative of Mickey's named Mortimer. This is not to mention immortal screen versions of characters like Dumbo, Bambi (Henry Ford's favorite), Peter Pan, The Three Little Pigs, Ferdinand the Bull, Panchito, Herbie the Love Bug, Maleficent the Evil Witch, Winnie the Pooh, Cinderella, Sleeping Beauty, Pinnochio, Brer Rabbit, Snow White, and, of course Dopey, Grumpy, Bashful, Sneezy, Happy, Sleepy and Doc. Tigers with cavities, pumpkins growing on the vine complete with jack o'lantern faces. Disney was probably the only artist ever to be praised by both the American Legion and the Young Communist League. His film with Donald Duck singing "We heil, we heil, right in the Fuehrer's face!" proved to be the most effective propaganda weapon of World War II, according to one writer; it was translated into ten languages and dropped behind enemy lines to be used by resistance groups. Mickey and his friends were responsible for Disney's winning thirty-one Oscars, four Emmys, honorary degrees from Harvard and Yale, the French Legion of Honor, and the presidential Medal of Freedom, more than nine hundred awards in all, not including the nomination by a leading French magazine in 1964 for the Nobel Prize. Pretty good for a country boy who once told reporters "I'm selling corn, I like corn!" And Disney always knew that he owed his success to the Mouse. He named the main street in his fifty-one acre Burbank studio "Mickey Mouse Avenue" in honor of his magical rodent. He hung a large oil painting of Mickey in his office next to smaller photographs and drawings of his human family.

No other animal, from Mary's little lamb and Moby Dick to Bucephalus and Lassie, no other animal factual or fictional, has ever captured the world's imagination the way Mickey Mouse did. Not even Barnum's Jumbo, the circus elephant whose name became a synonym for anything big. Mickey has always had his detractors, though. Not long ago, a talented, much respected author decreed that "one of the most disastrous cultural influences ever to hit America was Walt Disney's Mickey Mouse, that idiot optimist who each week marched forth in Technicolor against a battalion of cats, invariably humiliating them with one clever trick after another. I suppose the damage done to the American psyche by this foolish mouse will not be specified for another 50 years, but even now I place much of the blame for Vietnam on the bland attitudes sponsored by our cartoons."

A letter to the editor of the newspaper wherein the author's condemnation appeared in effect nolle prossed his case by suggesting that "a war-crimes trial be held *in absentia* for Mickey, Donald Duck, Porky Pig, and all the rest of that insidious crew, not excluding—let us not be squeamish about her sex—Snow White, plus the paramilitary group of dwarfs who used to march around behind her singing . . . and I humbly hope you might accept the presidency of the court."

Alas, there have always been Mouseophobes galore, but most share the opinion held by the eminent historian Marshall Fishwick. "Mickey's very existence," this great Mouseophile writes, "shows that the machine is capable, when properly directed, of producing heroes completely divorced from the biological process. This only increases our delight in seeing Mickey satirize the machine. . . . As with all heroes, Mickey's triumphs arise out of the desire of the human spirit to transcend the mechanical forces of brute nature. . . . Mickey's war on the machines makes him a symbolic foil of Frankenstein. . . . He releases us from the tyranny of things, and from the fear of the very intricacy and impersonality that makes him possible."

If there be any reason for the mousetraps that critics have set for Mickey, it is that within Disney's far-flung empire money often became the master, not the servant. But let others argue about Disney's triumphs and failings (Christopher Finch's *The Art of Walt Disney: From Mickey Mouse to the Magic Kingdoms* gives brilliant though somewhat unctuous praise; Richard Schickel's *The Disney Version* counterbalances his account). We're concerned only with the Mouse, which lives a life apart from his creator. Rather, the Mouse is that part of the creator that is pure simple-minded genius, perhaps almost the *imbécile de génie* that Count Keyserling called French novelist Georges Simenon. Mickey dates from the time in Disney's subconscious life when the cartoonist worshipped his retarded Uncle Ed, whom the family always looked after. Certainly there is a clue to his basically optimistic view of the world in his description of his favorite relative. "Uncle Ed may have been touched in the head," Disney once recalled, "but he was happy. He spent hours wandering through the woods. He knew all the birds and their calls and he knew the names of all the plants. It was a privilege to wander with him. . . . There wasn't anybody happier or friendlier and I loved him so. To me he represented fun in its simplest and purest form. . . . I could never figure out who was crazy, Uncle Ed or everybody else. . . ."

Probably everybody else. At any rate, Disney, the natural genius as assimilative in temperament as Shakespeare, the simple-minded man totally alien to intellectual analysis, best explained the phenomena of his and his Mouse's popularity. "Sometimes I've tried to figure out why Mickey appealed to the whole world," he once told a reporter. "So far as I know, nobody really has. He's a pretty nice fellow who never does anybody any harm, who gets into scrapes through no fault of his own, but always manages to come up grinning. Why Mickey's even been faithful to one girl, Minnie, all his life. Mickey is so simple and uncomplicated, so easy to understand that you can't help liking him."

There is no doubt that mice, particularly Mickey Mouse, embodied everything that Walt Disney liked, whereas Mickey's near equal in popularity, Donald Duck, said to be partly based on the Old Curmudgeon, Harold Ickes, was a combination of all the qualities Disney disliked in people. In any event, the esteemed British cartoonist David Low called Mickey's creator "the most significant figure in graphic arts since Leonardo Da Vinci." Similar opinions had previously been voiced by H. G. Wells, René Clair, and Thornton Wilder. In a tribute to Disney after his death in 1965, aged sixty-five, a *New York Times* editorial observed that what Disney "gave to us and the world . . . is all summed up in a friendly, engaging mouse named Mickey. It is not a small bequest." The editorial further noted that Disney had "a gift of imagination . . . somehow in tune with *everyone's* imagination," "a genius for innovation," "had transformed the simple story of comic animals into a universal language," but that "even more simply, he was the father of Mickey Mouse. . . ."

According to St. Louis artist Ernest Trova, Mickey Mouse, along with the Coca Cola bottle, the mushroom cloud, and the dread swastika, is the most powerful graphic image of the 20th century. We need only close our eyes to conjure up the Mouse, for he has been among the most pervasive visual images in our lives for the past fifty years. Kids started in the playpen with Mickey Mouse pajamas, Mickey Mouse cup, Mickey Mouse bowl, and Mickey Mouse spoon. Later, men at war, they were amazed to find an enemy cache uncovered near Hue in Vietnam that consisted of 300 Mickey Mouse sweatshirts—whether these were some secret weapon was never ascertained.

The Mouse mania continues unabated today at Mickey Mouse film retrospectives, parades and parties, and comic book conventions. Mickey has inspired many. Pop artist Roy Lichtenstein is reported to have adopted his comic strip style after his son challenged him to prove that he was a *real* artist by drawing Mickey, and artist Claes Oldenburg has based a series of sculptures and drawings on the regal rodent. Mickey has always been high art. Whatever the whole of *Fantasia* adds up to, the Mickey Mouse sequence in the film is inspired genius; in *The Band Concert,* Mickey conducts a William Tell Overture no other conductor has ever matched; similar praise could be bestowed on many of his adventures.

No doubt the Mouse has changed over the years, mellowed a bit too much, perhaps. At first the rodent and company were a marvelously vulgar bunch—"of the people," as one critic put it, unabashed, warm, alive and real, creatures of goodwill with no pretensions. But the studio gradually deprived Mickey of his rattishness. First the tail went, then the features were softened and his skin took on the pinkish glow of a contented Clarabelle. His face became cuter, his body, hands, and legs were more childish. His eyes, unfortunately, have been transformed from black ovals with a wedge-shaped hole to a conventional round shape with pupils, and even his short red pants now prudishly drop down to his shoes, while his girlfriend Minnie is never allowed to appear topless anymore. Perhaps this is the vapid mouse most critics take to task.

One art critic chronicles Mickey's progress or regression. Stephen Jay Gould, a Harvard University biologist, has recently stated in *The Panda's Thumb* (1980) that Mickey has been changed to exploit the human feeling of affection for the young. Mickey's change is an evolutionary phenomenon called neoteny, Dr. Gould says, a change that has been experienced by man as well because of its survival value: we retain to adulthood more juvenile features than our ancestors did. Dr. Gould's precise measurements show that Mickey's eye size has increased from 29 to 42 percent of head length; head length from 42.7 to 48.1 percent of body size; and that the change to a babylike cranium is even more pronounced. Writes author Amei Wallach: "From the wiseacre prankster of the '30s who was constructed in circles so perfect that animators sometimes used quarters to draw his head and body, and dimes for his ears, he took on a refinement.... By the '40s he had slimmed down and acquired wide white eyes to replace the art-deco ones.... By the '50s he was as bland and suburban as the country he symbolized. In the '60s he became a pop idol seen everywhere on watches and T-shirts. And now in the '80s, like the times, he has a complex, even a split personality. There is the Mickey of our childhoods riding the crest of the nostalgia wave, black eyes and all. And there is the streamlined new Mickey of the future... the successful host of Disneyland and Disney World."

Not even the staunchest Mouseketeer would have much good to say about the tiny (4′9″ to 5′3″) Mickey Mouse impersonators working at Disneyland, Disney World, and elsewhere (they are now available to promote supermarket openings and the like.). "You must never drink, smoke or indulge in crime or philandering as Mickey," they are told in their official guide's manual. And a sign in the Disney World staff cafeteria advises them to REMOVE YOUR HEADS (masks) AND PLACE ON TABLE AFTER ENTERING. The kids want no part of such human-size mice. More than one mother at the Mickey Mouse Festival in New York's Lincoln Center had to whisk screaming children away from these monster rodents. But plenty of bonafide Mickeys remain around; and no one has yet dared change the rodent's four-fingered gloves, his oversized shoes, or those black-disc ears that inspired the ubiquitous Mouseketeer hats. Nor is all wrong with the more mellow Mickey either; he has his champions, too. "There is of course more Christian feeling in late Mickey Mouse than in the 'Ave Maria,'" said the late John Gardner.

Recently, Mickey has made a kind of comeback and is emerging almost as popular as he ever was. In addition to being the symbol for such small universes as Disneyland, Disney World, and TV's Mickey Mouse Club, dolls in the Mouse's image are still the favorite with youngsters of all Disney toys and the recent fad for Mickey Mouse watches remains unabated after ten years. Not long ago the Mouse appeared in a special two-part musical comedy act at New York's Madison Square Garden called "Mickey's Revue," united again with such great stars of yesteryear as Goofy, Donald Duck, Jiminy Cricket, Clarabelle Cow, Horace Horsecollar, and Clara Cluck in the third edition of the revue "Disney on Parade." There he recreated his role in his classic cartoons "Orphans' Benefit" and "Mickey's Grand Opera." The show opened to excellent reviews, *New York Times* critic Howard Thompson noting that "It's hard to imagine any tot (or adult) demanding a refund."

The Mouse has had his legal troubles of late. Mickey even had to sue the Happy Hooker. It seems that a version of the familiar "Mickey Mouse March" was used as background music in a four-minute segment of a porno film called *The Life and Times of the Happy Hooker,* the march being played while three men wearing only Mouseketeer ears perform in an orgy with one woman wearing nothing. Disney lawyers submitted documents showing that the song, whose refrain "Mickey Mouse, Mickey Mouse" is familiar to millions, was written by Jimmie Dodd for Disney Productions and copyrighted in 1955. The suit sought $2.5 million in punitive damages, and a Manhattan Federal District judge issued a preliminary injunction barring the Mouse music from the sex film pending his final decision. But the damage was done. Asked one critic: "Can the Happy Hooker turn on without the tune of the Mickey Mouse March? Are Mickey's big round ears the newest measure of male sexuality?"

In another case the Mouse received a large number of write-in votes in a race against an unchallenged judge in Camal County, Texas. Some say he was clearly the most deserving candidate, but election officials refused to count his write-in votes on a technicality. Mickey Mouse might have won many elections in the past half century if it hadn't been for technicalities (thousands have voted for him over the years), but this time the technicality was downright insulting. According to a petition election officials filed: "Mickey Mouse is not and has not been a resident of Camal County for six months as required by law, and, furthermore, said Mickey Mouse is an idiot, lunatic and minor and very possibly an unpardoned felon and is therefore, under the laws of the state of Texas, ineligible to hold office."

Mickey Mouse has been most loved and kept alive by Mouse collectors, legions of them, who range from ordinary folk to artists like Andy Warhol and Ernest Trova. Trova owns 200 old Mickey Mouse watches and 100 boxes stuffed with MM comics and magazines, not to mention a wooden figure labeled "Micky" (without an "e") that is either the first Mickey Mouse toy ever made or a fake. Artist John Fawcett also has an enormous collection of mouseobilia and other "Mouse freaks" include partisans who can recite in order all the 140-plus Mickey movies. For collectors, there is even a book called *Disneyana,* filled with facts about the profitable hobby of collecting Disney treasures.

Collector Mel Birkrant, a New York toy designer, keeps a red, white, and blue "mouseoleum" in his living room; it includes a cast-iron Mickey Mouse bank and scores of wood and porcelain figures. But the champion Mickey collector is probably New Yorker Robert Lesser, whose collection has been valued at $5,000 and includes everything from radios and posters to Mickey Mouse pornographic booklets. These objects are carefully arranged in an East Side apartment where the bed sheets are printed with comic book strips. Lesser has a better apartment on Park Avenue, he says, but two nights a week he has to sleep in his fantasy world. The enthusiasm of such aficionados accounts for auctions like the one held in Los Angeles by Sotheby Parke Bernet where 170 items brought in a total of $14,885. It also explains the success of Finch's *The Art of Walt Disney: From Mickey Mouse to the Magic Kingdom,* which started out with a 20,000 copy printing at $45, 3,000 to 5,000 copies being the usual first edition order for expensive art books.

Yet Mickey never really needed a revival—he has already achieved immortality. Even when the decision was made three decades ago to bar Mickey from the screen because, as *Life* observed, he was "judged too sweet-tempered for current tastes," there emerged protest groups like the one calling itself "The Society to Keep Live Actors Out of Disney Films," in an age not at all noted for protest groups. "Neither Walt nor anyone else can deprive us of Mickey," Marshall Fishwick wrote. "Like Uncle Sam or John Bull, the Mouse has a separate existence off the drawing board. He belongs to those of us who accept him, as well as those who create him; he is public property."

Today the Mouse's name is a part of the language, adorning thousands of things, having appeared on some five thousand commercial products alone, from those ubiquitous $1.39 Mickey Mouse watches that now are collector's items (bringing up to $150 in good condition) to $3 sweatshirts, sheets, pocket calculators, and a diamond-studded Cartier Mickey Mouse locket selling for $1,150. In 1975 the Republic of San Marino issued a postage stamp honoring Mickey Mouse. Altogether the rodent has starred in over 140 movies, winning two Oscars, one in 1932 and the other in 1941 when he was busy selling war bonds. He has also been featured in four hundred funny-paper strips with a mass circulation of 48 million, and in books selling a total of 300 million copies. *Mickey Mousing* has become the name of a cartoon-sound track synchronizing process invented for the Mouse's films (each of the more than 100 ten-minute cartoons requiring 14,440 pictures, sixteen drawings needed to show a simple single step). "Mickey Mouse" was the password chosen by intelligence officers in planning the greatest invasion in the history of warfare: Normandy, 1944. *Mickey Mouse diagrams* were maps made for plotting positions of convoys and bombarding forces at Normandy. The special insulated boots issued to combat troops in Korea were called *Mickey Mouse boots,* and *Mickey Mouse discipline* was used and is still applied to childish rear-echelon inspections in the armed forces and elsewhere. The expression *Mickey Mouse* also means, among other things, anything that is trite or commercially slick in character, and *Mickey Mouse* in musical lingo means "a dainty high-pitched rendition." All and much more in honor of a *real* mouse who lived and died more than half a century ago, a real mouse reborn to revitalize fantasy and take his unassailable place among the great folk heroes of all time.

Perhaps this year—some year—a special Academy Award like that presented to Charlie Chaplin will be awarded to our rodent hero and the real Mickey (Mortimer) Mouse, or some relative, will skitter across the stage to the strains of "Hail to the Chief" or "Rule Britannia" and squeak his thanks to the thunderous applause of hundreds of less renowned celebrities, stars who unlike him have paid the toll of time but nevertheless are on hand to pay him homage.

Why not? "After all," as Professor Leo E. Litwak observed, "Walt Disney's history is in considerable measure our own. . . . For better or ill, at least one generation of Americans has Mickey Mouse in its bones. We don't have too much else in the way of a common experience. In a pluralistic society where our experiences of family, school and church are uncommon we are knit together by the Mouse. . . ."

# XI

# BILLION DOLLAR RODENTS
## Some of Your Best Friends Are Rats

Laboratory rats may see themselves as the most cruelly persecuted slaves in history, but as far as people are concerned the lab rat is the Hippocrates of ratdom; laboratory rats are to wild rats as Gandhi is to Hitler—they are a separate rat race of Koches, or Pasteurs or Salks, or Madame Curies. In truth we *have* conquered the rat in part by enslaving it. *Homo sapiens* may not be closely related to the rat, but 20 million rats are certainly our reluctant saviors and will in fact sacrifice their lives for us this year alone—many literally condemned to the gas chambers or guillotine. That is the estimated number of rats serving annually in U.S. laboratories alone; during your lifespan some 1.5 billion American rats will die for you, not counting the billions of non-American rats that will give up the ghost for humanity. Worldwide, more rats than Earth's entire present human population will die for us in the average lifetime of a *Homo sapiens.*

The laboratory rat has surely contributed more to the advancement of human health than any other animal in history, except his cousin the mouse (more about which shortly). Rats may already have saved your life, and have certainly helped you, if you've ever had a cold, suffered a serious injury, undergone surgery, taken prescription drugs for diabetes, heart trouble, high blood pressure and other chronic illnesses, or even had a vaccination. Research with rats has led to cures or progress in the battle against virtually every major medical problem from the common cold and air pollution to cancer, making the maligned rodents almost as important to human health as the scientist's microscope or test tube. "The fact is," as a National Institute of Health (NIMH) publication puts it, "that rats are indispensable to progress in medicine: in assisting medical scientists to discover the cause of disease, how illness affects tissues and organs, and what different approaches may work."

Rats and mice have contributed to breakthroughs in medical learning for several centuries now, long before 1877 when the relentless German country doctor Robert Koch injected untainted anthrax bacilli into a normal laboratory mouse, killing it, dissecting the creature, and proving for the first time in history that a specific germ could cause a specific disease, a discovery that led to cures for scores of the most dreaded human maladies. This hero rodent, perhaps the greatest animal hero in history, remains nameless, like most of the numberless noble rats and mice that have been sacrificed in scientific experiments at least since Robert Hooke and Robert Boyle used black rats for air cylinder tests in 1660.

Thumb through any index of scholarly scientific papers and within moments you will find hundreds of references to important experiments made with rats; well over 5000 articles about such experiments are published worldwide each *year.* The rat is especially valuable as a lab animal because of its similarity to humans. "Everything that happens in the human life takes place in the two years or so of the rat's life," says Dr. Freddy Homburger of Bio-Research Institute, Inc., in Cambridge, Massachusetts, one of the world's leading pure and applied research laboratories. "Every old-age change that you see in man—all the signs like arthritis, loss of hair, wrinkling of the skin, loss of hydration and fat, and so on—all that takes place in the aged rodent. . . . Rats also are subject to the same diseases of old age, and not only that, but to the same diseases at any time." Add to this the virtues that rats are relatively inexpensive to house and feed because of their small size, are just the right size to work with, are clean animals, and are fast breeders bearing large litters, and their value in the lab becomes even more apparent. They are equally valuable in medical, psychological, biochemical, pharmacological and ethological research.

"Given the power to create an ideal lab animal," says the venerable psychobiologist Dr. Curt P. Richter of Johns Hopkins University, "I could not possibly improve on the Norway rat." Rats have been much used in conditioning experiments employing the Skinner box and B. F. Skinner's vastly influential book *The Behavior of Organisms* is based on learning in rats. Rats are particularly useful in experiments involving intelligence because they have higher I.Q.s than the mouse and most other laboratory animals and learn how to solve problems faster. In one basic kind of learning experiment a rat learns by trial and error how to negotiate a maze and reach a reward of food. In another, the rat opens doors with a different design on each; it learns to distinguish a specific design on the door that has food behind it from the doors that do not. Such experiments give scientists insights into the basic steps of learning, although their conclusions must be carefully drawn because rats are not nearly as intelligent and complex as humans.

One recent experiment made on rats seemed to indicate that children who are able to exercise their minds should develop into brighter adults. Rats raised alone in cages containing just food and nests were not as intelligent when grown as were rats raised together in roomy cages with toys to play with and mazes to run through.

In another experiment, rats given electric shocks while in isolation tended to develop ulcers more often than rats in a communal cage who could release their aggressive reactions against each other, further evidence for the psychological theory that it is best for humans to release their aggressions in some way rather than keep them pent up inside.

The rat has also been used in many nutrition experiments, being especially appropriate for these because it requires a balanced diet just as humans do, except that rats don't need vitamin D. One study by the U.S. Department of Agriculture found that weanling rats fed yogurt had a higher growth rate than those fed whole milk, which may be the first hard evidence that yogurt is generally a nutritionally superior food. A second study showed that rats with a lower-calorie diet, two-thirds of what they normally ate, had longer life spans. Rats have frequently been utilized in controversial vitamin C studies, one of which has shown that if a lab rat were the size of a 150 pound human, it would manufacture between 1,800 and 4,100 milligrams of vitamin C on quiet days and upward of 10,000 mg. when ill or under stress.

Rats may even provide the answer as to why some people overeat and remain thin while others overeat and gain weight. After a 1982 experiment at Queen Elizabeth College in London, Dr. Michael Stock theorized that rats, and possibly people, possess a mechanism that enables them to burn off excess food energy as heat instead of storing it as fat.

Some lab rats have become alcoholics for science. When Dr. Gaylord Ellison provided a bar for rats in experiments at U.C.L.A., the rodents went on benders, suffered terrible hangovers and often became bona fide boozers who went on the wagon and fell off again within a few days. Alcoholic rats may one day provide answers to one of the major problems besetting mankind. In alcohol research, most extrapolations from rats have proved valid. Such experiments often require the use of specific organs and brain cells to determine the effect upon them and these must be obtained from alcoholic and nonalcoholic rats, as they obviously can't be taken from humans. Recently begun experiments may answer such questions as why chronic alcoholics develop a tolerance to alcohol, what makes chronic alcoholics so dependent on alcohol that without it they develop delirium tremens, and why alcohol tolerance carries with it tolerance to other drugs such as anesthetics. In alcohol studies, rats must be given quantities of booze larger than one might expect for their size, however, because they quickly metabolize the stuff.

In recent aging experiments, old rats injected with young rats' white blood cells have shown a greatly increased resistance to disease, which leads scientists to believe that someday they may be able to freeze a young person's white blood cells and use them when he or she ages, for a disease-free old age.

Sex studies may often be the ones that rats most enjoy. Not long ago, for example, a pill was tested that produced reversible infertility in rats. The animals maintained their normal sex desires and good general health, except for smaller testes. Though further research is needed, male birth control pills may result from the experiments. Perhaps less satisfied were those male rats used in an experiment which indicated that vasectomies, at least on rats, resulted in decreased hormone levels and more body fat.

While wild rats are dreaded plague spreaders, lab rats have given their lives in experiments with bubonic plague bacilli, as have lab scientists such as the microbiologist who lost his life at the top-secret Microbiological Research Establishment at Porton, England.

Some observers believe the rat is treated cruelly in labs, mainly because it is widely regarded as a "test tube" rather than a living creature. Rats are indeed commonly drowned, gassed, decapitated (there is a rat guillotine devised for these beheadings) and crushed to death for experimental studies; in some sex drive experiments male rats are allowed to mount females and then pulled off before they can ejaculate. Not a few scientists contend that the rat is too frequently used in lab tests, that in many cases other animals would be better suited. One psychologist quipped that the *Journal of Comparative and Physiological Psychology* should be called the *Journal of Rat Learning.*

Many experiments, of course, can't be done with rats or any rodents. Rats aren't used in arteriosclerosis experiments, though rabbits perform well in them. Rats cannot indicate they are nauseated, either, for they can't vomit, which is one reason poisons work so well on them. Each rodent species has its peculiarities, which make it appropriate or inappropriate for an experiment. Luckily, guinea pigs weren't used in the first penicillin experiments, for this worthy laboratory rodent is the only animal to which penicillin is poisonous and the lifesaving wonder drug might have been discarded if it had been tested on them. A species' peculiarities must be considered when testing it. In blood pressure experiments, for instance, scientists take a rat's pressure by wrapping the common inflatable cuff around its tail, but, in the words of one researcher, the rat's tail has the same relationship to the rat that the radiator does to the car: "So you have to warm him up beforehand in order to measure his blood pressure. You have to get him in a state where his thermostat allows the blood to flow freely through his tail. If it gets too cold, the blood just doesn't flow through it."

Space experiments have been made with rats and mice since at least 1951, when eleven mice were lifted 236,000 feet high by an American rocket and parachuted to earth, two of them dying in the process. Nine years later two rats and thirty-four mice were launched by the Russians and, along with two dogs and hundreds of flies, earned their place in history as the first living creatures ever to come back to earth from orbit in true outer space. Since then rats and mice have even been bred in outer space. The Russians made the most recent such experiment in late 1979, with U.S. participation, to determine if rodent offspring are affected by the lack of gravity. The rats were permitted to mate the second day in space and the orbiting laboratory landed before the female rats gave birth (in twenty-one days) because female rats might eat their young under the stress of space flight. When the rats returned to earth their offspring were to be examined at different stages of growth to look for abnormalities. The still-unpublished findings may help lay the groundwork for long human stays in space.

There are well over three hundred qualified breeders of rats and other laboratory animals in the United States today, according to a listing in *Animals for Research,* a publication of the National Academy of Sciences. Great care is taken in breeding the 100 million animals bred yearly for research, which was not the case as recently as forty years ago. "When I was a medical student in Switzerland in the late 1930s," Bio-Research's Dr. Homburger says, "we occasionally used small numbers of rats in certain experiments. . . . It was quite an experience to get these animals from the rat house, which was a dungeon-like windowless room in the basement of the institute where I was studying. One would don clumsy rubber boots, heavy protective gloves, and with huge metal tongs go hunting for the large animals that lived in the closely guarded room, roaming free amid an indiscribable mess of manure, food scraps, and newspaper bedding."

Although scientific breeding of rats began at Philadelphia's Wistar Institute in 1906 (one breed of rats is still called Wistars), modern animal breeding is an infant technology that can be traced back only to 1956, when the Charles River Breeding Laboratories of Wilmington, Massachusetts, discovered how to breed a disease-free rat. Charles River Laboratories had been founded by Boston-born veterinarian Henry Foster in 1947. Dr. Foster began by investing $1,300 in some rusty rat cages and today Charles River is the largest supplier of laboratory animals in the world, selling about 18 million animals a year for some $30 million. Customers include universities, government agencies, and manufacturers such as Dupont, General Foods, and Revlon, which use the animals for research and product testing.

Dr. Foster recalls that quality control in animal breeding was almost nonexistent when he began his business, the animals frequently shipped in fruit crates. All that came to an end when Charles River Laboratories pioneered in the breeding of disease-free rats, or what is called the "Caesarian originated, barrier sustained" (COBS) animal. Such animals are delivered at birth surgically and kept free of germs thereafter by precautions called barriers. All research animals do not have to be disease-free. Indeed, some experiments call for rats bred with a specific disease or disability, but those rat pups that are to be disease-free must first avoid contamination at birth in one of two ways, as a Charles River spokesman has explained:

> A pregnant female, at term, may either be sacrificed and the entire uterus (with liver intact) removed and transferred to the germ-free isolator via a germicidal trap, or a Caesarian operation may be performed on the mother. In the latter case, the female is secured to an operating table below, and is in contact with a plastic film in the floor of a sterile isolator. Using an electrocautery, the operator makes an incision through the adhered plastic sheet, skin, and body wall. The uterus is transferred to the security of the isolator and the pups delivered from the uterus. They are then transferred to a rearing isolator. In the former situation a hysterectomy is performed on the pregnant female, at term, and the entire uterus with the young inside is transferred through a germicidal trap to a flexible film isolator. Once the uterus is in the sterile chamber, the operator, with hands thrust into rubber gloves attached to the isolator, removes the young from the uterus and membranes and massages them gently to stimulate breathing. When this has been accomplished, the living pups are safely within the germ-free security of the isolator. The neonates [pups] may be nourished by periodic hand nursing or by foster suckling them on germ-free females.

The Caesarian-originated rats are elaborately protected for the rest of their lives. They are raised in quarters free from germs, animals, and insects, with temperature, humidity, air pressure, and air filtration precisely controlled. These contented rats, enjoying the seventy-five degrees Fahrenheit temperature and 50 percent humidity they like best, along with nonglare fluorescent lighting, eat specially prepared dried food that is sterilized and pasteurized before being delivered to their cages through pneumatic tubes. Sterilized bedding material is delivered to them the same way and the tubes are constantly used to remove waste from the cages. All personnel tending the rats must pass through three sanitizing rooms before reaching the animals, one in which they pass under an automatic insecticide, a second in which they remove their clothing, and a third where they perform a surgical scrub and don sterilized clothes and masks. Shipping of these animals to research labs is just as elaborate.

No wonder then that the rat has been called "a quality tool of the laboratory." There have been tests on rats over a three-year period that cost as much as $150,000, and the investment in labor, food, and overhead for *one* laboratory rat in an elaborate experiment can run as high as $10,000. Appreciation of the enslaved creature has recently been the subject of the full-length animated feature film, *The Secrets of the NIMH.* The movie cost a full $7 million to make and stars a group of experimental rats at the National Institute of Mental Health that as a result of injections secretly acquire humanlike intelligence and morality. Perhaps a greater tribute, and something of a rat thriller, came in 1938 when a German scientist at the Kaiser Wilhelm Foundation in Berlin fled the Nazis with two males and four females of a rat strain valued by geneticists for an hereditary quirk that makes 25 percent of the offspring wheeze and whistle when they breathe. He finally reached London, where the noted British scientist Professor J. B. S. Haldane relieved him of the whistling rats and sent them for safekeeping to America, where they thrived.

Only the rat's little relative the mouse has proved more valuable to scientists in the laboratory. Four of the five most valuable lab animals are rodents: the mouse, rat, hamster, and guinea pig. Even rodents like the huge Brazilian capybara are employed in laboratory research; this giant, weighing over one hundred pounds, is used in glandular growth regulation experiments. But the mighty little mouse is about as versatile as a laboratory animal can be. It has been called the most useful animal in the history of scientific research; about 40 million mice, twice the number of rats, are used in U.S. research alone each year. Like the rat, the mouse's only physical makeup closely resembles the human's, yet its smaller size makes it less expensive to use than the rat, if a little harder to handle.

Lab mice are still bred in the normal manner, by random breeding. But far more valuable is the inbred mouse, true inbreeding requiring the mating of brother and sister mice through a full twenty generations and producing mice with almost any specific quality scientists need. The desired trait can be a weakness, a low resistance to disease, a physical defect, or a special quality such as hairlessness. A hairless strain, for example, would be valuable in skin research, while female mice with a high rate of breast cancer would be valuable in cancer research. The genetic makeup of the individuals of any strain is identical.

University of California scientists Stephen D. Ferris, Richard D. Sage and Alan C. Wilson recently speculated that a sacred white mouse in the temple of Apollo thirty-two hundred years ago may be the single ancestress of at least nine varieties of inbred mice, numbering in the billions throughout the world today. By what amounted to fingerprinting the DNA (deoxyribonucleic acid) of mitochondria, cell structures that can be inherited only through the mother, they discovered the great mouse "grandmother" of the nine "old" inbred strains, whose vast numbers may be suggested by the 500,000 of one type, C57 Black, that the Jackson Laboratory in Bar Harbor, Maine, sold last year alone. "Our findings indicate that only one female lineage contributed to the formation of all the old inbred lines," the scientists reported. They suggested that the original "grandmother" mouse may have been one of the tame mice used by religious cults on Tenedos Island in the Eastern Aegean, where in 1500 B.C. the Greeks erected a temple to Apollo, god of mice, in gratitude to the rodents that had helped them win a great victory by gnawing through their enemy's bowstrings on the eve of battle. Priests in the temple thereafter bred the white mice for use in prophecy and, according to the California scientists, may have bred the white mouse ancestress, whose descendants were passed from priest to priest down the ages and finally to traders of the flourishing early European pet trade, who in turn sold them to scientific laboratories. If this is the case, there are billions of meek little mice in the world today with pedigrees four times as old as that of the most arrogant human noble.

Laboratory mice may prove to have the longest futures of any living mammal, as well as the longest pasts. About a century from now, hundreds of thousands of mice may be born of fathers and mothers that died in 1981. Five hundred or more frozen embryos from each of the over 750 mouse varieties or strains developed at Jackson Laboratory in Bar Harbor, Maine, will then be removed from liquid nitrogen refrigerator tanks, where they have been stored frozen in the laboratory at minus 196 degrees centigrade, and gestated in foster mothers until they are born as living mice for use in scientific experiments. Thus far mouse embryos frozen eight years have been successfully thawed, using a technique developed in 1972 at the Oak Ridge National Laboratory in Tennessee; but scientists see no reason why the embryos cannot be stored for centuries. Most of the varieties frozen are valuable because they have genetic quirks, such as tendencies toward certain diseases, that make them valuable in research work. Freezing them is safer, more convenient, and costs less than will breeding mice in the future. Initial storage of 500 embryos costs about $1,000 and the stock can be stored for $16 a year after that, whereas a living, breeding colony of mice costs about $500 a year to maintain.

The embryos to be frozen are taken from mice three days pregnant, when a mouse embryo consists of only eight cells and appears as only a tiny rosette under the microscope. After being examined for quality, they are sealed in plastic vials and carefully cooled in a liquid nitrogen tank until the temperature descends to minus 196 degrees. When it is ready to be used one or one hundred years later an embryo is carefully thawed and allowed to grow overnight in a tissue culture before being inserted into a foster mother, where it develops in the womb for another nineteen days. One out of four such transplants succeeds in producing mice today, but researchers are confident that this success rate will greatly improve as they gain experience in the technique.

Mice have been used in innumerable experiments covering almost everything imaginable. They have been employed in studies of the effects of new prescription drugs, mutation, transplants, the effects of pesticides, and skin grafting; every day hundreds of mice die so that man can find cures or relief for disease.

Mice were the first animals to test the possible dangers of material brought back from the moon. They have also been used in such unlikely experiments as the one in which University of Connecticut biologists succeeded in joining dormant teeth genes from chicken embryos with mouse cells—creating mice with chicken teeth! Then there was the mouse with eight parents produced at Yale University by Clement L. Markert, a leading authority on the cloning of animals, who merged individual eggs from different strains of mice at the earliest stage of development, these merged cells developing into a full-grown mouse.

Mouse calamities in laboratories have been much in the news recently. At a University of Southern California laboratory, for example, three thousand mice involved in a $1 million research project were killed when a computer malfunction caused lab temperatures to climb to over 100 degrees. Since the mice were being used in aging experiments, cumulative results were lost, delaying the experiments for as long as two years, with replacement mice for the experiment costing $60 to $80 each. In another similar case, eighteen months of research were ruined when an unknown person intentionally or accidentally left open all of an experiment's mouse cages at the University of Pennsylvania Medical School. Most of the 730 lab mice, which had been raised since birth for immunology and skin graft experiments, had to be destroyed because they were possibly no longer in the proper cages.

The worst mouse mix-up humanity knows about occurred in 1982 when tests indicated that some 120,000 inbred laboratory white mice of the widely used strain BALB/C were not genetically identical, as they were supposed to be. Hundreds, perhaps more than a thousand, of multimonth scientific experiments were ruined when it developed that the mice, often employed in cancer research and sold by one of the leading breeding labs, had over a period of fifteen months often been mislabeled and were not the BALB/C strain. Since all the mice of a specific inbred strain have identical genetic characteristics, scientists use them in tests where all the mice must be identical. For example, if a researcher wants to determine whether a certain drug prevents or cures cancer, he might give it to one group of mice and not to the other to see which group develops the most tumors. If the mice are not identical genetically, different genetic factors might influence tumor development, regardless of the drug's effect.

Some scientists had to throw out the results of over a year's painstaking work because of the mix-up, which could easily have ruined five-hundred years' worth of experiments that might have saved hundreds of lives if they had been completed on schedule. The supplier, which experienced similar trouble with both mice and rats in the past, has since destroyed the contaminated stock and begun a genetic monitoring program, but the problem is a growing one worldwide, having been reported in Europe and Japan as well in recent months. Human error is usually to blame in such situations, contamination occurring when animals from different strains are inadvertently mixed, or escape from their cages and mate to contaminate the pure gene pool. One expert, writing in a journal of the National Academy of Sciences, estimates that "4 to 20 percent" of all laboratory mice throughout the world have become genetially contaminated, a figure that involves millions of mice and no one knows how many human lives.

Yet, despite their alarming frequency, recent contamination incidents in breeding laboratories are clearly the exception to the rule. For the most part pure, clean rats and mice are provided to scientific researchers; these rodents are humanely treated and given tender loving care. Says one laboratory director: "We immediately fire anyone found intentionally causing an animal pain or even teasing or taunting animals." The people who abuse lab animals are indeed a very small minority today, for humanitarian as well as economic reasons. The predominant attitude is summed up by Dr. John Foster of the Center for Disease Control, who declares, "When you deprive an animal of its freedom, you accept the responsibility for its welfare." Adds another scientist: "It's little enough to be kind for what sometimes amounts to only a few days to rats or mice who are responsible for adding so many years to human lives."

# XII

# PET RATS

Hollywood animal trainer Moe DiSesso must be the rat keeper par excellence among petowners and may well understand rodent psychology better than anyone since the Pied Piper. DiSesso is the man who trained the five hundred or so rats that starred in the horror films *Willard* (1971) and its sequel *Ben* (1972). Starting with six female and six male rats he allowed them to reproduce, and began his training program. The first problem was to teach rats that humans meant food and security and could be trusted. That DiSesso has the utmost confidence in his solution is obvious when one considers that he had his teenage son lie down in the rat cage fifteen times a day so the rats would get used to people. After a while the rats trusted their new cagemate, who apparently trusted them, too, since they began crawling all over him. Soon the swarm of rats were trained to bare their teeth and squeak on cue, enter and exit from a suitcase, and climb stairs and work their way up and down furniture. Baby rats were taught to identify the sound of an electric buzzer with food by being rewarded with tastes of peanut butter. DiSesso even taught his rats to take instructions and work together in groups of twenty-five. The scenes where they "attacked" human beings were made possible by smearing the actors' bodies with peanut butter.

DiSesso was hardly the first person to train rats or make pets of them. Rats and mice probably were caught and tamed centuries before seventeenth-century Japanese nobles began breeding the "fancy" inbred varieties that some science historians say were the ancestors of those invaluable laboratory strains that we use in experiments today. There is little doubt that people caught and made pets of house rats and mice when they first attached themselves to man's grain stores in Asia ten thousand years ago. They have been enjoyed as pets ever since, and by a surprising number of well-known people. There have been vaudeville acts of trained rats performing tricks like pulling small wagons and catching fleas. Picasso and Walt Disney had pet mice, as mentioned keeping their charges, respectively, in a desk drawer and an inverted waste paper basket. Today all animal experts agree that both rats and mice make excellent, safe pets that are easy to care for. As far back as 1898, C. A. Conklin, an animal handler who ran the zoo in New York's Central Park, assured us that women were no longer afraid of rats, that that was an ancient and invalid stereotype. "Talk about women and mice," he said, "its women and *rats* nowadays. Yes, you wouldn't believe it, would you? But it's the truth, there are a number of women in New York who have rats in their homes for pets. I don't know whether it is what you call the 'new woman' movement or what, but there certainly is a change in women."

Perhaps we have vanquished the rat in a small way by taming and making a pet of it, but rats never will take to the lap the way the cat does or take to the leash the way the dog does, though cartoonist Dwaine B. Tinsley did a penetrating cartoon of kids and their pets at an "Annual Children's Pet Show" where the judges are reviewing an obviously middle-class Gerald and his sheep dog "Ruby," his female counterpart Rachel and her cat "Fluffy," and finally an inner-city child, Billy, who holds his pet house rat "Ignacio" on a leash.

By no means try to catch a wild house rat (or house mouse) for a pet. Not only might you be bitten and infected, if you do succeed in capturing your prey (and they can be caught alive in certain traps). These wild rodents are invariably infested with parasites, many of which will unhesitatingly adopt a new host: you. They also carry various kinds of viruses, bacteria, fungi, and protozoa that may be pathogenic for humans. Large house rats are especially fierce when cornered or caught, even cats seldom attack them, and house mice have an objectionable "mousy" odor that domesticated strains do not exude. If you must raise wild house rats or mice for a scientific experiment that demands their use, proper precautions should be taken to avoid infestation and infection; consultation with biologists experienced in handling such animals should be sought. But as a rule stay away from wild rodents. One animal dealer specializing in reptiles told this rat writer that he wouldn't even feed a wild rodent to his snakes for fear of contamination, despite the fact that wild rodents are much cheaper than domesticated types.

On the other hand, strong bonds can form between individual wild rats and people, as happened in the case of rodentologist Dr. Stephen C. Frantz, who had to release into the wild a bandicoot rat he had tamed while in India. A week later the rat returned when Dr. Frantz called it with a clucking sound to which the rat always responded: "Out staggered this battered rat. The wild rats had really worked him over. He came up to me, sat on my foot and wouldn't get off. He had had enough. . . . I couldn't take him back, but I never felt easy about leaving him there. I still remember him and wonder what happened to him. I don't like rats in general, and yet with some individuals you establish a relationship. I mean, they are *life,* aren't they? They are *living* things."

Some experts do recommend various species of wild field and wood mice as pets. These wild mice don't possess the objectionable mousy odor of the common house forms and can be interesting, attractive pets if properly cared for. The common, widely distributed white-footed or deer mouse (*Peromyscus maniculatus*) is most often suggested as a wild rodent pet. Found almost exclusively in woodland areas, it is distinctly different from other mice; its large ears, big eyes, white undersides, and pinkish white feet make it look like a kind of Mickey Mouse in evening clothes. White-footed mice, which rarely bite, can be handled with impunity right after capture and are tame enough to feed from the hand twenty-four hours later. They are exceptionally clean; a newborn white-foot, in fact, will wash its face after each meal, making this a practice even before its eyes open. Baby white-foots are carried hanging onto their mother's breasts. These musical mice are a pleasure to hear, chirping away like birds, but be warned that they can be plague carriers and check with local health authorities before making a pet of one. *The same applies to any wild rodent mentioned here.*

Other wild mice that do well in captivity include the jumping mouse (*Zapus hudsonicus*), which can leap as far as nine inches when excited, and its close relative the western harvest mouse, which likes to use its long prehensile tail to hang from overhead positions much like a monkey. The southwestern seed-eating pocket-mouse (*Perognathus fasciatus*), so named for its external fur-lined cheek-pockets, also makes an excellent pet. Most interesting and useful is the western grasshopper mouse (*Onychomys leucogaster*), an attractive mouse that eats insects. Zoologist Clifford B. Moore, director of the Forest Park Museum in Springfield, Massachusetts, used to let a pair of grasshopper mice run free in his New York apartment; within three weeks the savage, fearless little creatures had rid the apartment of cockroaches!

All wild rodents, even the meat eaters, will adapt to an ordinary mouse diet of grains, eating whatever you would feed domesticated rats and mice. Zoologist Clarence F. Smith, writing in the *Journal of Mammalogy,* noted that the harvest mouse prefers wild oats, brome grass, foxtail, burr clover, and white-stem filaree. Emerson C. Brown, former director of the Philadelphia Zoo, concocted a wild rodent diet of 20 parts cracked oats, 10 parts buckwheat, 5 parts cracked corn, 5 parts whole wheat, 2½ parts sunflower seed, 1½ parts millet, and 1 part peanuts, served each day with lettuce and fresh water. Most wild rodents can be housed like their domesticated brothers and sisters, except for the jumping kinds. To fully appreciate jumpers, provide them with plenty of room: cages at least four feet square and two feet high.

Far easier to obtain than wild rodents, and generally more preferable as pets, are the domesticated rats and mice that have been bred commercially for nearly a century. The Carworth Farms CF#1 strain mice are especially placid and easy to work with. There are more than 100 different strains of such mice to choose from, each with differing genetic makeups that affect behavior. Most pet shops carry a good selection and, as noted, some 300 sources of high quality laboratory rodents are listed in the annual *Animals for Research* published by the National Academy of Sciences. Among rats the gentlest and easiest to care for are the Sprague-Dawley, Wistar White, and White Animal Farm strains. The Wistar strain of piebald rats, started by inbreeding a wild female born with a white chest spot, and the Long-Evans strain of black-and-white "hooded" rats are a little more difficult to handle but worth the extra trouble. Other choices are the Irish rat, black with white feet; the wild agouti, with banded colors; and tan, red and blue rats. Mouse strains include the Angora, hairless, fat, white, yellow, silver, red, and spotted kinds. Especially interesting are the waltzing mice, which can move only in circles due to an inner ear defect affecting their balance; they frequently go into an extended wild spin, sometimes an entire family of them "waltzing" at once. Their nesting compartments should be provided with front and back doors so that they can enter spinning through one and exit spinning through the other!

Bear in mind that while female mice are odorless, males have a distinct mousy smell; buy a "buck" only if you want to breed mice, keeping it with two or three does. More than one male in a cage usually means trouble. Females begin breeding at the age of two months and give birth to litters averaging 5 to 8 in number within 19 to 21 days, and can have as many as 17 litters or 135 offspring in a year. Remember that if every one of these mice were allowed to breed and proved fertile, over thirty thousand mice would result in a single year, considerably more than you'd want to house or feed.

Mice should be handled carefully because they are more nervous than rats and will not hesitate to bite when startled, or nip when they smell food on your hand. There are two basic methods of handling them, methods which can be applied to rats as well. For mice accustomed to people, gently using a cupped hand is satisfactory. The towel method should be used for frightened or aggressive mice. Just hold a small thick unfolded towel in one hand and gently but securely grasp the mouse behind its head with the same hand, supporting its body underneath with your other hand. The same method can be employed by wearing gloves in place of the towel. Mice that are difficult to catch can be grabbed by the *base* of the tail between fingers and thumb and then scooped up with the other hand. Don't grab the *end* of the tail; in some mice a protective reaction causes the skin of the tail to come off when they pull away after the tail is grasped. Any slight bite

The deer mouse is sometimes kept as a pet today. *(Richard Lydekker, The New Natural History)*

Hamsters, popular pets, in their natural state *(Richard Lydekker, The New Natural History)*

incurred while handling a mouse or rat should be thoroughly washed and disinfected and an antitetanus shot may be needed for a deep bite, especially from a rat. There is, however, very little danger of infection from a well-cared-for mouse or rat.

Mice should never be kept in direct sunlight; as little as fifteen minutes of strong light is enough to kill them. They are nocturnal creatures, but one way to enjoy them when they are most active, without staying up all night, is by keeping them in a dark place during the day and lighting their cage after sundown. Cages should be especially secure, for mice, unlike rats, often will not return to their cages if they escape. Mouse cages should be at least eight by ten inches in area and six inches high. Mice do as well in a wire-topped aquarium made for fish as they do in commercially available plastic and metal cages; wooden cages are harder to keep odor free, because the wood absorbs urine. If a wooden box is used, always cover the inside wood surface with wire to prevent the mice from gnawing their way out. It is a good idea to provide all cages with an exercise or metabolism wheel. Mice also like to run on elevated runways, climb ramp ladders, and swing from trapezes.

Keep the floors of mouse cages completely covered with a bed of pine wood shavings, sawdust, shredded paper, dried rice hulls, dried corncob pieces, or commercial litter to a height of about one inch. Eucalyptus oil (one teaspoonful to a quart of water) makes an excellent deodorizer when spread with an atomizer, but the litter should be changed at least every other day to eliminate the strong accumulated odor of urine, as mice urinate almost constantly. Mites and lice that sometimes infest mice can be controlled by using one part cedar wood shavings to four parts pine shavings as litter. The cage itself should be thoroughly cleaned with soap and water every ten to fourteen days and at cleaning time the mice should not be allowed to run free but kept in a holding cage until their permanent home is ready. It should be noted that, according to a recent study, the strongest-smelling mouse cages are those that are cleaned *daily* because after each cleaning the males mark out their territory with a strong-smelling substance; they do not feel the need to repeat this performance until the cage is cleaned again.

Mouse and rat quarters must be kept at a temperature of from seventy to eighty degrees Fahrenheit, free of drafts and with a relative humidity of about 50 percent. It is very important that no chloroform be used in the room where mice are kept; they are extremely sensitive to chloroform vapor, which destroys their kidney tubules and causes them to die of kidney failure. Fresh drinking water should always be available. Rather than keep a water dish in the cage use a six-ounce water bottle with a one-hole rubber stopper that is fitted with a stainless steel or glass drinking tube, through the end of which both mice and rats will drink.

Mice eat very little, averaging about one-seventh of an ounce of food a day, compared to the cage rat's ounce or so a day, and they enjoy a variety of foods. Commercial preparations are fine, providing all their basic needs, but they can be fed varied diets including raw vegetables, nuts and fruit (go easy on the last to prevent loose stools), corn, grains, cereals such as granola or rolled oats, birdseed, sunflower seeds, dog food, dog biscuits, bread crusts, boiled egg, and even cat chow. Cheese, strangely enough, is not very good for mice and meat should be limited as a steady meat diet tends to make them vicious and cannibalistic. They should be fed twice a day, in the morning and midafternoon, and when food is left over from one meal, the following meal should be slightly reduced to avoid overfeeding. The advantages of commercial food pellets are that they insure a balanced meal for the animals and are easier to work with. The pellets can be put into a wire mesh feedbox that is suspended from the top or side of the cage; the rodents eat the pellets through the mesh. This avoids fecal contamination of the food. Pellets can be kept in the cage at all times, but other foods should be removed the same day if not eaten.

Pet mice will live as long as three or four years (more in exceptional cases), growing from 1.5 grams at birth to 30 grams at maturity, whereas a pet rat will live the same average three to four years and grow from 5.5 grams at birth to about 300 grams (500 grams for a male). Both begin breeding in as early as forty days, having a natural heat cycle every four to five days, and a healthy breeder female can be expected to produce about forty young before she stops breeding at the age of about one year. Mice and rats are quite easy to sex, as the testicles and penis on the males are very evident. They especially like having a lonely nest site above the ground in the cage, or a separate breeding cage, but provide them with plenty of warm nesting material such as shredded newspaper or cotton from which they can fashion their nests, and make sure the cage isn't brightly lit. It is a good idea in fact, to provide the cage with a small nesting box.

Observe the mother mouse or rat and her hairless, pink, blind pups at birth, but take care not to disturb them and never handle them during the first week. Mice and rats may kill and cannibalize their babies out of anxiety over interference, though this is unlikely except in extreme situations. The fragile pups begin to crawl out and into the nest in about two weeks and leave their mothers after a month or so. Up to this time the mother will respond to their squeaks of alarm and retrieve them when danger threatens as a cat does with her kittens. Remember to separate the males from females when they leave the nest and put them into different cages, maintaining the same ratio of a male and two or three females in each eight by ten-inch cage if you are going to keep them all. Experts recommend that defective pups be humanely destroyed with a couple of drops of chloroform on a piece of cotton. Overcrowding in a cage not only disturbs rodents but causes tumors in mice.

It is interesting to note that female mice usually do not go into heat unless a male mouse is present. Another intriguing phenomenon is the Bruce Effect: if a pregnant mouse is exposed to a strange male during the first four days after mating with another male, her pregnancy will be blocked and she will bear no offspring. And although mice can become pregnant just after delivering a litter, such a pregnancy (called delayed implantation) will last twenty-five to thirty days instead of the usual nineteen to twenty-one days, so that she can finish nursing her first litter before having a new one.

Mice and rats are clean animals, frequently grooming themselves and each other; bathing them is not necessary and could be dangerous to their health. Neither are they prone to many serious illnesses when properly cared for. Pet mice sometimes contract mouse pox (ectromelia), Tyzzer's disease, and salmonellosis, but these diseases are rare. Pet rats are sometimes subject to chronic respiratory diseases. Rather than trying to diagnose the many diseases these rodents can, but rarely do, have (middle-ear infection that causes head tilt; chronic diarrhea; crust around the nose caused by bacterial infection; viral eye infections; gangrene of the tail caused by low humidity; body swellings caused by tumors; and runny noses caused by pneumonitis), keep a careful eye on your pet and catch illnesses in their early stages. Remove unhealthy animals from the cage until they can be checked and treated by your veterinarian, keeping them in a "sick cage." You can treat pet mice and rats yourself for minor scratches with a diluted iodine or three-percent solution of hydrogen peroxide, and dust them with flea powder to control mites and lice. Be sure to dust according to the directions on the package of pyrethrin powder or rotenone, which shouldn't be used for young animals.

Rats are raised essentially like mice except that the larger rodents (a rat usually grows ten times the size of a mouse) should be kept in bigger cages. A two-foot square wire cage eighteen inches high makes a good home for five to six rats in captivity, though a plastic cage or covered aquarium is fine, too. It is a good idea to have some wood in the cage for the rat to gnaw on; without this their front incisor teeth grow so long that they get out of line and have to be trimmed. Rats are more intelligent than mice, as smart as a dog or cat, actually, and make better pets because they are easier to handle, bite less, do not smell as bad, and unlike mice, they grow slowly, making them more interesting to watch. Rats also exhibit a large variety of body language display, ranging from their crouch for grooming to the threatening posture they take when challenging another rat. As "breeding factories" mice are preferable to watch, but rats seem to have more individual "personality."

Rats that are let out of their cages almost always return voluntarily because they know food awaits them. Mice are more difficult to retrieve. Never anxiously rush after any escaped rodent, however. Try to lure it back, with a bit of food if the animal is out of reach under a piece of furniture, and let it come into your hands. Patience is the keyword in taming or training mice and rats, too. You must be gentle with the animals; in fact, commercial laboratories use professional "gentlers" who do nothing but handle rats daily and keep them so tame that they can be handled with no danger of a bite. Move slowly and speak in a low voice. Lure the rodent onto your hand by putting food onto your palm and any time the animal seems frightened let it go. The tricks it learns will come out of its store of instinctive behavior if you provide ladders, wheels, and other props and reward an activity with special food treats.

Rats and mice can be used in many interesting home experiments. Several can be made by constructing a maze out of plywood or sheet metal and testing the animals in it. Make the maze by tracing it out on a sheet of paper and then nailing down the pieces of wood or sheet metal to form the pathways. You can, for one example, find out whether frequently handled rats or mice solve the maze faster than those that are not handled. Or determine how the age of mice or rats affects their learning ability. Or discover whether food-deprived rats or mice find the food at the end of a maze faster than well-fed specimens. The possibilities are endless.

If you really enjoy raising rats and mice as pets, your hobby might be turned into a profitable business. Laboratory supply houses and biological supply houses (their addresses can be obtained from reference guides in any public library) were, at last report, paying $1.60 each for a healthy rat and twenty-five cents each for a healthy mouse. Pet shops are another possible source. Write several of these companies inquiring about their standards and prices before you begin a rat or mouse farm, and be sure to read more on the subject of large-scale rodent breeding. Success stories are plentiful in this area. About twenty-five years ago, for instance, a husband and wife team in Rochester, New York, began a small garage-based mouse-and-rat-breeding concern that is today the multi-million dollar Blue Spruce Farms.

# XIII

# A RAT COOKBOOK
## Or Eating Well Is the Best Revenge

People have been eating rats, and enjoying them, much longer than many people would care to admit, or even contemplate. "A man's palate can, in time, become accustomed to anything," Napoleon said while exiled on Elba, and in his *Consuming Passions* Peter Farb remarked that humans will gulp down anything that doesn't swallow them first. Very strong preferences indeed have been shown for things like flies in honey sauce, sheep eyes, bowels of bluejay, vulture intestines, cockroaches, worms, ants, mushrooms soaked in urine, baked pubic hairs, and different kinds of dirt, not to mention some of our favorite pets such as the dog, cat, and horse. Despite various religious restrictions, our species has generally been very practical about eating other species, and even our own species sometimes. We seem to proceed with the philosophy of the Englishman, V. M. Holt, who published a little book in 1885 called *Why Not Eat Insects?* "The insects eat up every blessed green thing that do grow and us farmers starve," Holt reasoned. "Well, eat *them* and grow fat!"

Since ancient times the Chinese have regarded the rat as a delicacy almost equal to the cat and the dog, which are even today selected live from cages wheeled to the table in certain Hong Kong restaurants. The early Chinese relished rats, called them "household deer," in fact, and often hunted both the common rat and the bamboo rat. Recurrent famine has eradicated most food taboos in south China and even rat fetuses have been highly regarded in the local cuisine. Going a step further, or backward, depending on your point of view, Chinese in Lingnam during the T'ang dynasty (618–907) ate newborn rats stuffed with honey; the animals "crawled about the banquet table peeping feebly" and were snatched up with chopsticks by eager diners to be eaten alive. A Chinese scholar seeks to explain or apologize for this as "apparently an aboriginal habit adopted by Chinese settlers."

In 759, "when Kuo Tzu-i besieged An Ch'eng Hsu in Yeh-ch'eng, the price of a single rat rose to 4,000 cash," another Chinese scholar tells us. But there are some stories proving that it could be taboo to eat rats in China. One old folk tale, for example, has a rat falling into the bird's nest soup and ruining it.

The strongest ancient restriction regarding rats is found in the Old Testament, where in Leviticus, chapter 11, and Deuteronomy, chapter 14, the basis for many Jewish dietary practices, the consumption of any "creeping animals such as the mouse," is forbidden. Since there is but one Hebrew word for rat and mouse this reference could be to the fat dormouse, the fat sand rat, or the jerboa, all of which were eaten at the time in the Middle East. In any case, according to Biblical dietary laws, anyone who even touches a "swarming thing that crawls upon the earth" is "unclean until evening" and any clean thing that falls upon such a swarming thing "must be put into water, and it shall be unclean until the evening." When Christianity was born and broke with Hebrew dietary proscriptions, for various reasons, the prejudice against the rat as food was of course a taboo never abandoned by most converts. One does not need religious reasons to abominate the rat.

Christianity, however, never placed a restriction on the eating of rat flesh and neither apparently has any other major religion save Judaism. The Buddha, for example, banned the eating of elephants, horses, tigers, lions, panthers, bears, hyenas, serpents, and humans, but not of rats. Many primitive tribes did forbid consumption of the common house rat, though, because it consumes excrement or garbage, this being an adaptive step that prevents human contact with parasites. According to one anthropologist, members of a New Guinea tribe still eat any of twenty-one rat species, but not the house rat. The Maoris, nevertheless, introduced the Polynesian rat to Polynesia as a food in the sixteenth century or earlier, carrying it with them in their ships as a living food supply.

The ship rat is merely a house rat on vacation or migrating to the promised land, but that hasn't stopped sailors from dining on it since earliest times. Shipwrecked sailors have several times compared rat meat to terrapin in taste and, indeed, at the turn of the century a well-known French restaurant was accused of substituting rat meat for terrapin; there was supposed to be an albino rattery where the "terrapin" was raised. Chinese restaurants seem to be accused of substituting rat meat for terrapin and similar dishes more often than other ethnic eateries in our time, just as they are accused of using cat and dog meat. Proof of such accusations is totally lacking.

Antonio Pigaffeta, the only survivor to write an account of Magellan's voyage around the world, told of the Patagonians' eating "mice raw without even slaying them." Whether this is worse than the rats and sawdust Magellan's starving men were themselves eventually forced to eat is a matter of opinion, but it is a fact that his crew were forced to dine on rodents. This practice saved many from scurvy, as it did aboard the celebrated Arctic exploration ship *Advance* centuries later. Toward the end of the disastrous voyage, on which Magellan lost his life in a skirmish with natives, rats sold at a ducat each. Pigaffeta's memoirs didn't mention whether the rats were enjoyed or not, but the French seaman-explorer Louis Antoine de Bougainville wrote in his journal aboard the *Boudeusa* that "at supper we ate some rats and found them good."

In *Rats, Lice and History,* Zinsser tells us that at the French garrison of Malta during the rebellion of 1798, "food was so scarce that a rat carcass brought a high price." Similarly, during the Siege of Paris in 1871, when Prussian troops besieged the city for 135 days, Frenchmen ate labeled rat meat sold at the rat market in the Place de l'Hôtel, the rodents costing three francs apiece (as compared to ten francs for cats) and dishes like *salmis de rats à la Parisienne* (a *salmis* is a dish made mostly with game, two-thirds cooked) could be had in restaurants for about fifteen francs. The National Guard hunted rats, most of which were sold to the meat markets. To Parisians who survived the terrible siege, having "eaten rat" became the symbol of their greatest humiliation. And many ate rat without knowing it. So many falsified foods were on the market at the time (wolf was even sold as lamb!) that no one could be sure what he or she was eating. The noted chef Thomas Genin, who died in 1887, told how rat was

Gustave Doré engraving of rats being sold for food in besieged Paris, 1871 *(Radio Times Hulton Picture Library)*

substituted for rabbit and other dishes. "The rat was repulsive to the touch," he wrote, "but its flesh was of tremendous quality: delicate but not too insipid. Well seasoned it is perfect. I have served grilled rats as *pigeons à la crapauderie* (trussed pigeons prepared in a certain way), but more often as potted meat, with a stuffing of donkey's meat and fat."

During the Siege, in which even sick animals from the Paris zoo were eaten, one Frenchman reputedly "fattened up a huge cat, which he meant to serve up on Christmas day, surrounded with mice, like sausages." These mice were almost certainly ordinary house mice, not the famed fat or edible dormouse (*Glis glis*) of the Romans, which has long been relished in Europe. The fat or edible dormouse is doubtless the most celebrated of rodents served at the table, the Roman poet Martial even having written an epigram about the wee beastie:

> I sleep all the winter and become fatter
> During that time in which nothing but sleep finds me.

Martial is referring to the fact that the fat dormouse more than doubles its weight during autumn in preparation for its long six-month winter hibernation. The Romans used to catch this little vegetarian when it was at its fattest, about 240 grams, roast it, and dip it into honey. Dormice were also stuffed with a mixture of pork, pine nuts (one of their favorite foods), pepper, asafetida, garum sauce and broth, sewn up and baked on a tile in an oven.

The Roman historian A. Marcellinus complained that pretentious Romans set scales upon the table so that the fat dormice they proudly served could be weighed by notaries, who informed the guest just how extraordinarily fat were the delicacies their host set out as fare. According to the historian Varro, the Romans actually raised the sleepy rodents in specially designed breeding areas called *gliarii.* Varro provided a description of these:

> The glarium must be completely surrounded by a wall made of smooth stone or smooth mortar on the inner wall, so that the animals will not climb out. Inside the pen should be bushes, those that grow nuts. If these bushes do not bear any fruit, then the animals must be fed acorns and chestnuts. In addition, holes for these animals to bear their young in must be made. They do not need much water, as they drink only a little, and are accustomed to living in arid areas. The fat dormice are fattened up in barrellike pots like those in country houses; however, the pots used for fattening up these animals are made quite differently from the others. The animals make small pathways and a small hollow in the wall where they store their food. One feeds these animals large amounts of acorns, chestnuts, or other nuts. Consequently, they become fat. The barrellike pot is kept dark.

The famed Roman gourmet Apicius, who is said to have committed suicide when he ran out of enough money to support himself in his usual opulent style (he only had a few million remaining), left behind a recipe for stuffed dormice that is considered a classic but far from exact. Instructed the great chef: "Take Dormice in pork fat, with bits of meat taken from every limb of the dormouse and pounded: stuff with pepper, kernels, asafetida, sauce; place them in tiled receptacles, put them in a warm place or cook them in an oven."

In what is now Yugoslavia during the early nineteenth century the dormouse was considered a delicacy by rich and poor alike, its fat highly regarded as both butter and a fat for frying. It is still roasted, broiled and fried with crackling in southeastern Europe, Austrians finding it a nutlike delicacy, and the French prizing it as a "tender almond-flavored" gourmet dish. That particular taste is ironic, because an almond-flavored liqueur as well as an almond-flavored biscuit is called *ratafia* or *ratafee.* As recently as 1972 an enterprising Englishman proposed raising fat dormice for food, claiming that his countrymen would come to love the mouse meat as much as did the Romans who once ruled Britannia.

Besides eating dormice for pleasure, the Romans used them in various home remedies. We have the sometimes questionable word of Pliny that dormice ashes mixed with honey will cure an earache, and, as we have seen, he recommended that infants be given "sodden mice" to eat if they wet their beds. In the Middle Ages eating a fried dormouse was believed to cure smallpox, though eating food any mouse had nibbled was thought to cause a sore throat. Centuries later an old Pennsylvania Dutch recipe recommended ordinary house mouse pie as a cure for bed-wetting, measles, and whooping cough, while in our own time mouse milk has been tested as an aphrodisiac, albeit an expensive one, for the milk, used in laboratory experiments, costs $10,000 a quart.

Frank Buckland, perhaps the strangest of an odd array of nineteenth-century English naturalists, as a boy learned how to cook field mice and specialized even then in mice baked in batter. "A roast field mouse—not a house mouse—is a splendid *bonne bouche* for a hungry boy," he later pronounced. "It eats like a lark." Buckland, whose home was a virtual zoo, including a rat that could sing and a cellar full of tame mice (he had no compunctions about eating his pets), became a confirmed zoophagist and sampled as many different animals and birds as he could get his teeth on, from earwigs to elephant trunks. But like father like son. Buckland's father, the dean of Westminster, had raised the boy on dinners that featured mice, not to mention other delicacies such as crocodile for breakfast and puppies for lunch. Dean Buckland once claimed that he had eaten the heart of king Louis XIV of France which had been preserved by Lord Harcourt. He even sampled bat urine more than once, licking it off a marble floor. Compared to such provender, boiled, fried or broiled mouse seems positively mouth watering.

Field mice are still eaten in several countries, including Mexico, where they are skinned, eviscerated, skewered, and roasted over an open fire. In the Arctic, at least one explorer has recently marinated mice in ethyl alcohol for about two hours, dipped them in flour, fried them for about five minutes, simmered them in alcohol for another fifteen minutes with several cloves, and served them in a cream sauce.

Rodents were eaten so frequently in Frank Buckland's time, or were so often considered as a source of food, that an anonymous wit managed to get the following fake report, from a spurious natural history council, published in the unsuspecting *Gardener's Chronicle:* "In birds a great success has been obtained by the Hon. Grantley Berkeley, who has succeeded in producing a hybrid between his celebrated Pintail Drake and a Thames Rat; and the Council considers that this great success alone entitles them to the everlasting gratitude of their countrymen, as this hybrid, both from peculiarity of form and delicacy of flavour (which partakes strongly of the maternal parent) is entirely unique."

It is definitely no hoax to state that rats and mice are an important source of nutrition and widely eaten today. Some people eat them unknowingly. At least one American, as noted, got a rat in his Coca Cola. Upton Sinclair became a vegetarian for many years after writing *The Jungle* and discovering, in his investigation of meat-packing plants, that rats were often sold as beef. In 1933, a Prague butcher shop was raided on the complaint of a customer, and police found "many hundreds of smoked rats bodies" hanging from nails in the back room. The butcher said he had "for some time been doing a good trade in smoked rat's flesh" and that his customers from the very poorest class in the population were "grateful for the chance of buying meat so cheaply." Officials said the rats had been carefully prepared and weren't dangerous to human health, police charging the man only with operating without a butcher's license.

Rats sold for twenty francs apiece in Brussels during the early days of World War II. Inmates of Nazi concentration camps often had to eat rats to survive, as did the defenders of Leningrad, and some thirty years later the Viet Cong reportedly ate rats in Vietnam. On what may be a lighter note, Watergate's Gordon Liddy has written that when about ten years old he killed, roasted, and ate a rat to prove to himself that he could do it.

Mr. Liddy did nothing that thousands, perhaps millions of less macho people do today. Rats are, in fact, a dietary staple in West Africa, where they comprise more than 50 percent of the locally produced meat in parts of Ghana and are a regular part of the diet in eastern Nigeria. Not long ago a scientist, trying to establish which rat species carried deadly Lassa fever, showed a photo of a native rat to a group of African villagers and one man cried out impulsively, "Oh, that's delicious!"

Between 1968 and 1970 over twenty-five thousand pounds of greater cane rat (*Thryonomys*) meat were sold in just one African market in Accra. The greater cane rat, called a "ground pig," is the object of organized hunts that take place almost daily during the dry season in many parts of that continent. The reeds are fired and the rats driven into the open, where the natives make a sport of killing them similar to the pigsticking the British enjoyed in India, except that the Africans are on foot. Dogs are also used to round up the rats and drive them into areas where they can be conveniently slaughtered with spears and clubs. Once alerted, however, the rats are hard to catch. In one clever maneuver a rat will rush at a dog, stop suddenly and remain motionless right under its adversary's nose. The dog, startled and confused, often turns tail and runs.

A favorite local recipe in Ghana calls for skinning and eviscerating the cane rat and splitting it lengthwise. It is then salted, browned in peanut oil, covered with tomatoes, water, and hot peppers, and simmered until tender. The giant-tailed rat and giant-pouched rat are prepared in similar dishes. Such fare has been hailed as a way to vanquish *kwashiorkor,* the terrible protein-deficiency disease that plagues Africa and has killed so many children.

Among the many other people who eat rats for protein are New Guineans, who consider the flesh of the foot-long mosaic-tailed rat (*Uromys*) a great delicacy and climb up to hunt it while it sleeps during the day in the fronds of palm trees. Thais in rural provinces particularly enjoy the rice rat (*Oryzomys*), similar to American rice rats, which are killed on hunts twenty thousand at a time, prepared like chicken or rabbit, and sold for as little as the equivalent of thirty-five cents a pound.

Filipinos take culinary revenge on a subspecies of the dreaded black or roof rat (*Rattus rattus*), whose ancestors carried the Black Death bacillus to Europe centuries ago and which eat a full half of their rice crop now. The rats are killed with bare hands, sticks, and sickles while the rice stalks are being harvested; a small group of harvesters often accounts for a thousand rats in a single day. Skinned and gutted, these rats are dried in the sun and either taken home for supper or sold in local markets for five centavos or so apiece. Deep-fried, they are served in steaming coconut oil and are said to taste pleasantly gamy like squirrel or rabbit. About ten years ago a local businessman in Los Banos in the Philippines tried canning rat meat, giving it the trade name STAR, "rats" spelled backward. It didn't catch on, but plans were already in the works for a local rat sausage.

Close rat relations that people value as food are legion. Porklike porcupine meat is enjoyed from Italy to South Africa; muskrat meat, said to taste like chicken, is widely marketed as "marsh rabbit" or *musquash,* its American Indian name; and hamster meat is relished by many. A trapper who ate ground squirrel in 1906 started an epidemic of plague in San Francisco, as did Chinese hunters who ate infected tarbagan meat four years later. As recently as 1977 a Navajo Indian boy was stricken with bubonic plague after preparing and eating prairie dog, which is a Navajo delicacy when baked in mud.

South Americans have long eaten rats and other rodents, centuries before beef, pork, and or poultry were commonplace foods. Elizabeth Wing, a zoo-archeologist at the Florida State Museum, says that guinea pigs, which are little more than oversized rats, were hunted and trapped in the Andes as early as 10,000 B.C. By 300 B.C. they were domesticated, becoming the first native animals raised for slaughter in the Western Hemisphere. Guinea pig is still domesticated and eaten in South America. Guinea-pig feed is in fact sold in local markets next to stands selling the roasted rodents themselves. Locally, guinea pig is prepared much like suckling pig. The rodent is scalded in hot water, its hair removed, and its skin scraped with a knife before it is cleaned and roasted. According to the World Health Organization, some 7 million guinea pigs are "harvested" annually as food in Peru. It is edifying to note that, in a 1975 symposium held at the Rockefeller Foundation, Dr. James McGinnis of Washington State University suggested guinea pig as a protein source (its flesh is about 19 percent protein, about the same as beef, pork, or poultry) because the rodents can utilize the fibrous plant material cellulose, which man cannot digest.

Rodents are so popular south of the border that Barbara Ford warned in *Future Food* (1977): "If you've eaten a dish in South America and don't recognize the meat, it may well be a giant rodent." The paca, the cavy, the agouti, the pacarona, and the capybara are all South American rodents hunted for their flesh. The semiaquatic capybara (*Hydrochoerus hydrochaeris*), the world's largest rodent at up to 174 pounds, is second in favor only to the guinea pig, and has long been considered a good food. While Darwin tasted its flesh during his 1832 voyage around South America and found it only "indifferent," an anonymous contributor to the British magazine *All Year Around* in 1861 told readers that the capybara "strongly tempts the domesticator, feeding on water weeds and thus converting into wholesome nutrients vegetable substances which are turned to no account. It is very prolific and produces a great quantity of meat in a short space of time." "Anonymous" proved most visionary. South American natives do still hunt the capybara in canoes, killing it when it comes up for air. They make ornaments of its large teeth and, of course, eat its flesh. But most capybara are raised on ranches along with cattle today, grazing side by side with the cows and happily consuming less vegetation than the cattle for the meat they yield. On one ranch in Venezuela, where capybara is considered a Lenten food because people believe (or want to believe) that it is a kind of fish, a total of 1,500 were slaughtered in one year. South American scientists Juhanie Ojasti and Gonzalo Padilla told the North American Wildlife and Natural Resources Conference in 1972 that the capybara might be an excellent food source and suggested marketing it as smoked sausages or as a canned meat to "widen its appeal."

Nutrition expert Nelson Chaves urged his government to solve Brazil's chronic food shortage by encouraging poor people to eat the common house and field rat. Chaves was subjected to some kidding—*Esquire* magazine listed his suggestion among the "Dubious Accomplishments of 1980," for example—but, as we've seen, he was hardly proposing anything new. It is of course true that much care must be taken in preparing rat meat to guard against deadly and debilitating diseases, and that rat meat may be no panacea for the world food problem. But eating rat flesh would certainly help solve that problem and the flesh itself may be tasty, judging by the fact that the Chinese and French cuisines, among the greatest in the world, give rat meat a place today. Modern Chinese, like their ancestors, find rats tastier than pork, salting and sun-drying them for use in many recipes, some of which are highly praised as hair restoratives! Contemporary Frenchmen liken their taste to partridge or pork and in the famed cookbook *Larousse Gastronomique* there is indeed a recipe for them: *Entrecôte à la bordelaise* (Grilled Rat Bordeaux Style). "Rat," according to one French gourmet, "has a slightly musky taste that is not unpleasant," though it has to be submitted "to prolonged cooking to destroy the harmful germs." Rodents from the wine-making Girónde district are captured in the wine cellars there, carefully cleaned, singed, and boiled. They are then skinned, eviscerated, their tails cut off, and brushed with a light layer of olive oil and coarsely chopped shallots. The rats are supposed to be especially good grilled over old vineshoots or broken wine barrels.

"Few of us are adventurous in the matter of food," said the American philosopher William James; "in fact, most of us think there is something disgusting in a bill of fare to which we are unused." However, a major characteristic of cuisine the world over has been its willingness to experiment, and as Brillat-Savarin said, in *The Physiology of Taste,* "The discovery of a new dish is more beneficial to humanity than the discovery of a new star." Perhaps this will be the case with rat in the near future throughout the world. At the very least we might accomplish what the English poet laureate Robert Southey suggested when he predicted that man would defeat the destructive rat only when he made him a table delicacy. Bear in mind, however, two things:

That rats are inefficient in converting vegetation to meat, so that no matter how many we partake of we may still be fighting or eating a losing battle.

And that Brillat-Savarin also said, "Tell me what you eat, and I will tell you what you are."

# APPENDIX I
## Great Ratborne Plagues in Recorded History*

1491 B.C.—Biblical plague
1300 B.C.—Egypt
378 B.C.—Rome
68—Rome
125—Rome
164—Rome
767—Worldwide
453—Rome
430—Athens
187—Egypt, Syria, and Greece
167—Roman Empire
169—Roman Empire
189—Roman Empire
250–65—Roman Empire
336—Syria
381—Antioch
410—Rome
430—Great Britain
434—Italy

---

*This list includes a large percentage of the 109 rat-caused plagues estimated by some scholars to have struck in the first 1500 years since the birth of Christ and the 45 bubonic plague epidemics from A.D. 1500 to A.D. 1720. Many are more fully described in the text. Several plagues of uncertain diseases are included, where there is a chance that they may have been bubonic. All plagues listed from 1333 to 1720 are considered to be part of the Black Death pandemic.

444—Great Britain
531—Byzantium
Constantinople)
542—Roman Empire
746-49—Constantinople
774—Scotland
863—Scotland
1008—Wales
1068—England
1078—Constantinople
1094-95—London
1095—Ireland
1097—Palestine and Egypt
1106—England
1111—London
1123-24—France and Germany
1172—Ireland
1175—England
1204—Ireland
1218—Damietta, Egypt
1235—England
1262—Ireland
1271—Ireland
1333-37—China
1340—Italy, 2d Pandemic
1348-1666—Europe
1361-62—France and England
1367—France and England
1370—Ireland
1386—Smolensk, Russia
1407—London, England
1466—Ireland
1470—Dublin, Ireland
1471—Oxford, England
1499-1500—London, England
1563—London, England
1600—Russia
1601-03—Ireland
1603-04—England
1611—Constantinople
1625—England
1632—France
1656—Italy

1664—London, England
1672—Naples, Italy
1672—Lyons, France
1711—Austria and Germany
1720—Marseille
1740—Messina
1760—Syria
1770—Poland, Russia
1773—Bassora, Persia
1792—Egypt
1799—Africa
1850—Kwangyung, China, 3d Pandemic
1894—Canton
1896—Hong Kong
1898—Turkestan
1900-01—Australia
1900—San Francisco, California
1901—Hong Kong, China
1901—Egypt
1902—Egypt
1903—India
1903—Punjab, India
1903—Niuchwang, China
1904—India
1905—India
1906—India
1907—India
1907—Seattle, Washington
1909—Tuantsiu, China
1910-11—Manchuria
1910-13—China and India
1914—New Orleans, Louisiana
1921-23—China and India
1924—Los Angeles, California
1930—Tunisia
1933—Manchuria
1935—Uganda
1940—China
1948-60—Cases in many areas
1963-83—Vietnam, cases in many areas

# APPENDIX II
## Other Dangerous Ratborne Diseases

As noted in Chapter II, rats carry 35 or more often fatal diseases, including Bang's disease, bubonic plague, Chagas' disease, dysentery, foot and mouth disease, Lassa fever, listeriosis, lymphocytic choriomeningitis (LCM), mud fever, Newcastle disease, rabies, rat-bite fever, relapsing fever, Rickettsialpox, Rocky Mountain spotted fever, Salmonellosis, scrub typhus, splenic fever, swine fever, toxoplasmosis, trench fever, trichinosis, typhus, tuleremia, and Weill's disease. Following are brief accounts of a number not described in the text.

*Lassa Fever.* The newest disease shown to be carried by the rat is African Lassa Fever, which produces fevers up to 107 degrees and swift death. American medical workers invited to investigate a serious outbreak of this disease by Sierra Leone discovered that the deadly virus causing it was carried by *Mastomys natalensis,* a small 3- to 5-inch brownish-colored house rat unique to Africa. Transmission of the disease could occur either through contamination by rat feces or urine, animal bites, or by eating the rat, which is a common food source in many parts of Africa. Three scientists died in making the discovery.

*LCM.* A disease of mice and rats that causes many of the symptoms of encephalitis, and sometimes death, is viral lymphocytic choriomeningitis, or *LCM. LCM* infects humans through contact with the saliva, nasal secretions, urine and feces of infected rodents, usually in contaminated food or dust. The disease is interesting to scientists because of the curious tolerance of congenitally infected mice to it; the mother passes *LCM* directly to the embryos in her uterus and the babies are quite normal except that an injection of blood from one of them will kill a mouse from clean stock.

*Rabies.* Rabies from rat bite remains a problem in certain areas, although in 1969 the U.S. Public Health Service advised that the bites of rats or mice seldom call for rabies prophylaxis in the United States, noting that "this recommendation, based on the low rate of infection in wild rodents, is further supported by the fact that there has never been a case of human rabies *in this country* attributed to rodent exposure, even though rodent bites are common." All rodent bites should be promptly cleaned and disinfected, however, and tetanus shots or boosters are often recommended.

*Rat-bite fever.* The rat transmits two types of ratbite fever. The type most common in the United States is Haverhill or streptobacillary fever, this acute disease caused by streptobacillus moniliformis, an organism inhabiting the teeth, gums, nose, and throats of rats. One to five days after being bitten and contacting Haverhill fever, the victim develops chills, fever, skin rash, and severe joint and back pains. The related spirillary rat-bite fever or *sodoku* of the Orient develops in 5-28 days and often subsides only to reappear again a few days later.

*Rickettsealpox.* No scientist, but a New York exterminator discovered the cause of Rickettsialpox, a disease occurring so often in the Kew Gardens section of New York City during 1946 that it was first called Kew Gardens Typhus Fever. The disease should rightfully have been named for Charles Pomerantz, head of a local exterminating firm, when he established the cause of it while investigating the basement of a modern apartment house where the pestilence seemed to have originated. Pomerantz found that the mice in the building were infected with the house mouse mite (*Leponyssoides sanguineus*), which killed the mice with the disease and then left their furry havens to crawl into warm rugs and beds where they bit humans and infected them with a milder nonfatal disease. Pomerantz-pox would be more appropriate than "Rickettsialpox," and more alliterative, too.

*Salmonellosis.* This disease is named after Daniel E. Salmon, pathologist and veterinarian of the federal Bureau of Animal Industry, who first identified the rod-shaped bacteria that cause the disease, which is an acute gastroenteritis that can be spread in various ways, especially through food contaminated with rat feces containing *Salmonella* organisms. There are often severe outbreaks of the disease, such food poisonings almost as common as those carried by staphylococci.

*Scrub Typhus.* The mouse, as well as the rat, carries scrub typhus, a *Rickettsia* transmitted by the blood-sucking larval stages of certain lice that the rat hosts. Often called by its Japanese name of *tsutsugamushi* fever, scrub fever is a severe disease with a mortality rate as high as 50 percent. During World War II in the southern Pacific it ranked next to malaria as the most important cause of disability and death from disease, some 7,000 cases and 284 deaths reported by American forces alone.

*Trichinosis.* Rats play a major role in the spread of trichinosis, a major health problem throughout the world that results from an infestation of the intestines and muscles by larvae and cysts of *Trichinella spiralis.* Rodents and man both develop trichinosis from eating insufficiently cooked pork infected with the organism. Rats play an important part in the spread of the disease to hogs feeding on uncooked garbage at dumps where trichina-infested feces of rats are present, or where rats are caught and eaten by pigs. The rats in turn, often feast on infested pork scraps at dumps, which keeps the rodent-pig-man cycle of the disease going.

*Tuleremia.* Infected wild rodents can infect humans with tuleremia, once called a "purely American disease" because it was first scientifically observed in the United States, named after Tulare County, California, where it was found, and because Americans first discovered the causative agent, *Bacterium tularense,* and described its transmission from one species of animal to another and to man. Its most severe form is deadly tularemic pneumonia.

*Weil's Disease.* Weil's Disease, or leptospirosis, is often a severe infection that causes death in humans who come in contact with infected urine of rats containing the spiral-shaped bacterium causing the diseases. Excreted into polluted water the corkscrewlike spirochettes can last relatively long periods and enter the body through mucous membranes or small cuts and abrasions in the skin. Weil's disease, also called yellow jaundice, is frequently found in sailors, swimmers, fishermen, miners, sewer workers, firemen (when they pump contaminated water), fish and poultry dealers, and others who work in wet places where rats congregate. There have been many outbreaks of the disease in the United States, where thirty percent of all rats carry it. Not long ago, extensive outbreaks occurred among Australian sugarcane cutters working in wet fields contaminated by rats.

# APPENDIX III
## A Rat Poison Primer

# SOME CHARACTERISTIC

| POISONS | | Lethal Dose (Mg/Kg) | Percent Used in Bait | Degree of Effectiveness | Acceptance | Reacceptance | Cumulative | Tolerance Developed | Odor | Taste |
|---|---|---|---|---|---|---|---|---|---|---|
| **ANTICOAGULANTS**<br>**Warfarin**<br>**Fumarin**<br>**Pival** | •• | 1[1] | .025 | Good | Good | Good | Yes | No | None | Slight |
| **ANTICOAGULANT**<br>**Diphacinone**<br>**Chlorophacinone** | •• | 0.5[1] | .005 | Good | Good | Good | Yes | No | None | Slight |
| **ANTU** | • | 8[2] | 1.5 | Good | Good | Poor | No | Yes | Slight | Medium |
| **FLUOROACETAMIDE (1081)** | •• | 15—Norway<br>51—Mice | 2.0 | Good | Good | Good | No | No | None | Slight |
| **NORBORMIDE** | • | 12 | 1.0 | Good | Fair | Poor | No | Yes | None | Slight |
| **RED SQUILL** | • | 500[3] | 10.0 | Fair | Fair | Poor | No | No | Medium | Strong[4] |
| **SODIUM FLUOROACETATE (1080)** | •• | 5—Norway<br>2—Roof R.<br>10—Mice | 1/2 Oz./Gal.<br>1 Oz./28 Lbs. | Good | Good | Good | No | No | None | Slight |
| **STRYCHNINE (Alkaloid)** | ••• | 6 | 0.6 | Fair | Fair | Poor | No | Yes | None | Strong[4] |
| **STRYCHNINE (Sulfate)** | ••• | 8 | 0.8 | Fair | Fair | Poor | No | Yes | None | Strong[4] |
| **VACOR** | •• | 5—Norway<br>18—Roof R.<br>98—Mice | 2.0 | Good | Good | Good | No | No | None | ? |
| **ZINC PHOSPHIDE** | •• | 40 | 1.0 | Good | Good | Good | No | No | Strong | Strong |

• Effective against Norway Rats only.
•• Effective against Norway Rats, Roof Rats, and House Mice.
••• Mice only.

1. More or less. Successive doses required for 5-10 days or more.
2. Norway rats only, on first exposure.

# COMMON RODENTICIDES

| ity | Type of Bait Mixtures | | | ACTION (Cause of Death) | Relation to Humans and to Other Animals | | | ANTIDOTES |
|---|---|---|---|---|---|---|---|---|
| Oil | Dry | Fresh | Water | | Secondary Poisoning | Absorbed thru Skin | Degree of Hazard in Use | |
| Yes | Yes | No | Yes | Inhibits clotting of blood; causes internal hemorrhages. ■ | Rare | No | Slight | Vitamin K and transfusions of whole blood. |
| Yes | Yes | No | No | Inhibits clotting of blood; causes internal hemorrhages. ■ | Rare | No | Slight | Vitamin K and transfusions of whole blood. |
| No | Yes | Yes | No | Pleural effusion (over-production of fluid in the lungs). ■ | No | No | Medium | None |
| No | No | Yes | Yes | Similiar to sodium fluoroacetate (1080). ■■ | Yes | No[5] | Extreme | None |
| No | Yes | Yes | No | Blood vessels constrict, causing failure of organ systems. ■■ | No | No | Slight | None |
| Yes | Yes | Yes | Yes | Heart paralysis. ■ | No | No | Slight | Acts as own emetic to animals capable of vomiting. |
| No | No | Yes | Yes | Paralysis of heart and the central nervous system. ■■ | Yes | No[5] | Extreme | NONE. Monoacetin or ethyl alcohol and acetic acid recommended. |
| No | Yes | No | No | Convulsions due to super-stimulation of nervous system; exhaustion; asphyxia. ■■■ | No | No | Medium | No emetic after 10 minutes. Charcoal in water and sedative. Keep in dark room. |
| No | Yes | No | No | Convulsions due to super-stimulation of nervous system; exhaustion; asphyxia. ■■■ | No | No | Medium | No emetic after 10 minutes. Charcoal in water and sedative. Keep in dark room. |
| No | Yes | No | No | Respiratory failure. ■■ | No | No? | Medium | Nicotinamide has been shown to be antidotal in lab rats. |
| Yes | Yes | Yes | No | Heart paralysis; Gastro-intestinal and liver damage. ■■ | No | No | Medium | Copper sulfate before emetic; Cathartic and water. Avoid fats and oils (as milk). |

■ Slow acting
■■ Fast acting
■■■ Very fast acting

3. Minimum acceptable level; more toxic squills give better results.
4. Normally objectionable to rats.
5. Can be taken through cuts or breaks in the skin; also danger of inhaling loose powder.
6. Emetics used as first aid except as noted; speed is essential; 1 tablespoon of salt in a glass of warm water is usually effective; call a physician immediately.

d from
pt. of Interior, U.S. Fish and Wildlife Service Leaflet #337, Revised Dec. 1959.

# APPENDIX IV
## Rat Riddance Recipes
### A Universal Illustrated Guide for Rat Detectives

The first logical step in getting rid of rats is to learn how to look for rats. The most positive proof is, of course, seeing a live rat, but rats are secretive animals and are usually visible only when there is a large infestation. One way to check is to switch on the lights in a darkened suspect room and listen for scampering. Rat sounds, heard only when an area is otherwise quiet, include running, gnawing, and scratching. Various squeaks and churring or chittering

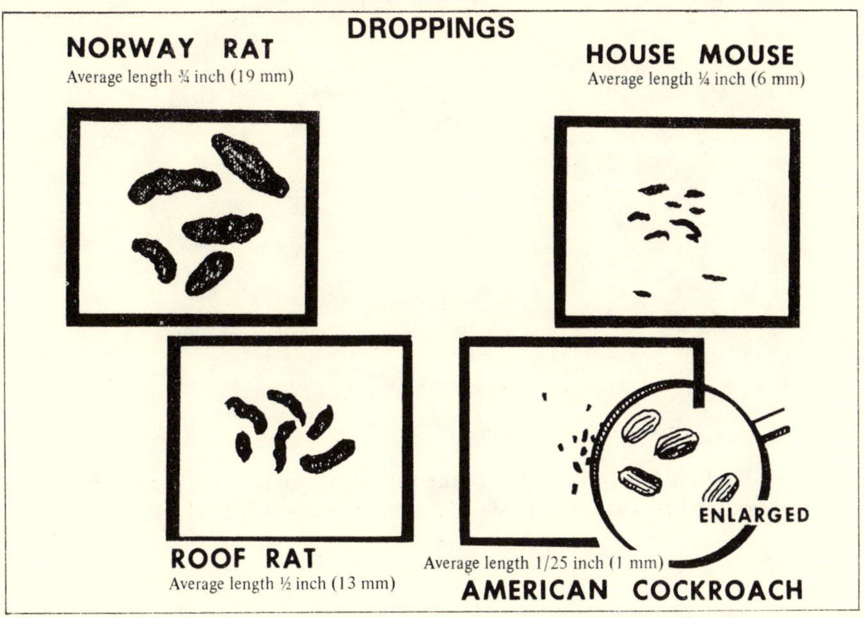


*(U.S. Public Health Service)*

noises may also be heard intermittently for several minutes and rat pups in the nest make faint squeaking sounds.

Fresh droppings are a sure sign of rats and they can indicate the type of rodent that has infested an area. The droppings are usually black and they are shiny and moist when deposited within a few days, dull black or even grayish and hard when any older. Variation of sizes of the droppings indicates that rats of several ages are present and that reproduction is probably occurring in the colony. Rat droppings are mostly found along runways, behind objects near walls, in secluded corners and near food supplies. They are not found in burrows or nests, which are usually very clean, rats having been observed carrying feces from them.

Runways, rubmarks, and tracks are also reliable signs of rats. Creatures of habit, rats almost always use the same runways between food, water, and their harborage. Outside, their runways are narrow paths of beaten earth swept clean of debris; indoors they are found along walls, steps, and rafters. Rats prefer continual body contact with at least one vertical surface, such as a wall, because

The dirty "swing marks" of the black or roof rat on overhead beams indicate its runway *(U.S. Public Health Service)*

of the keenly developed sense of touch in their whiskers (vibrissae) and specialized body hairs and thus a dark, greasy rubmark forms along regularly traveled runways from contact with the rat's body. When fresh, these rubmarks are soft and will smear if touched. Norway rat runmarks are usually found at ground level while roof or black rat rubmarks are seen as overhead swing marks beneath beams or rafters at the points where they connect to walls. Rat tracks along runways and in other areas are sharp and distinct when new, the imprints of the five-toed rear paws more commonly seen than those of the four-toed front paws. Dusting suspected runways with talcum powder or flour helps one check for fresh rat prints and tail marks. The hind feet leave prints up to 1½ inches long. Some experts believe that long, wavy marks left by a dragging tail indicates old rats.

Look for signs of rat gnawing around doors, windows, utility lines, and packaged goods, especially in food storage areas. When rat gnawings are new they are light-colored, show distinct tooth marks, and have small chips of wood or other material near them.

Rat burrows are easy to find, occurring along the outside walls of buildings and outbuildings, in dirt basements, and in earth banks, rubbish piles, hedge-

Rat guards on a ship's hawsers *(U.S. Public Health Service)*

rows, and under heavy growths of brush or bushes. Burrows are usually close to a source of food or water. The entrance holes average approximately three inches in diameter and holes in recent use are beaten down, free of dust and cobwebs, and often have fresh earth pushed out of the entrance in a fan-shaped pattern. The burrow systems are shallow, but can be quite complex with several ratholes leading to the same system.

Rat nests are usually too well hidden to be worth looking for. Neither are urine stains or hairs, both hard to identify, reliable signs of rats. When there is a heavy rat infestation, an unmistakeable, peculiar, musty odor is present, especially in poorly ventilated areas.

Ideally, sanitation for rat control should be practiced long before rats have been located in or near a dwelling or outbuilding. "If you don't feed them, you won't breed them," is the slogan one pest control agency uses, but such measures will help even if used after rats are spotted and certainly after they have been spotted and eliminated, or reduced, by poisoning or trapping (see chapter 6). There are three essentials for good sanitation (1) eliminate the sources of rat food, (2) remove rodent shelters, and (3) ratproof the premises.

Starve rats out by keeping garbage and refuse in tightly covered containers (the first earthenware pots were made not for decoration but to keep foodstuffs safe from rats and mice). Also store food in ratproof buildings, rooms, closets, or containers, and keep the entire premises clean and food free, down to the burners on the stove. To help keep rats from biting children, always wash toddlers' faces of milk and take away their bottles when they fall asleep. Incidentally, good methods of defense against any attacking rat (should you be pinned down or

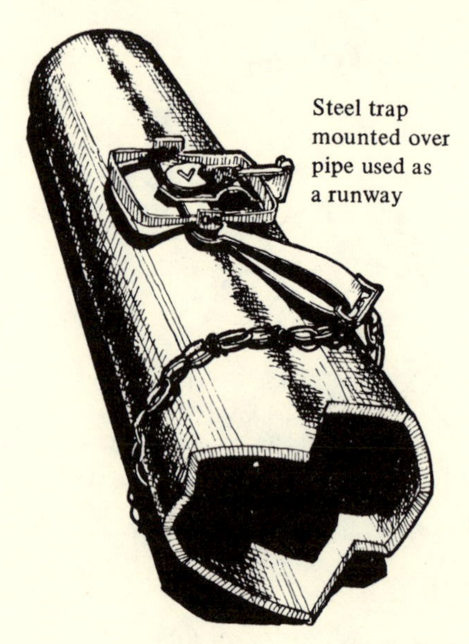


A clever way to catch a rat
*(U.S. Public Health Service)*

otherwise helpless) include blowing on it (rats hate this), screaming at a high pitch, and even biting it, according to one authority, if you can gather up the courage.

Remove rat shelters by concreting all basement walls and floors, storing materials in the basement on stands one foot above the floor, and not allowing litter to accumulate on the floor behind sink, stove, or cabinet. Outside the house do not pile wood, coal, or other objects near or against walls; concrete all wooden steps and keep stored food twelve to eighteen inches off the ground.

Buildings can be ratproof and rats have never crossed the threshold of modernday structures like the Empire State Building. Remember when ratproofing any area that a rat can squeeze through an opening the size of a quarter dollar or a half square inch. Be sure all doors fit their frames properly, leaving no space for rats to enter, and use good springs to make sure that doors close properly. Vents and windows can be made secure by screening them with heavy wire mesh in a sheet-metal frame. In addition, floor drains should be tightly fastened to stop rat entry from sewers, and openings around pipes should either be filled with concrete or patched with sheet metal. It is also a good idea to place metal guards around or over wires and pipes to prevent rats from using them to gain entrance to a building.

*And if nothing else works, there is always that Welsh charm to write down and slip into the mouth of "King Rat":*

r.a.t.s.
a.r.s.t.
t.s.r.a.
s.t.a.r.

# GENERAL BIBLIOGRAPHY


Barkley, A. *Studies in the Art of Ratcatching.* 1891.

Barnett, S. A. *The Rat, A Study in Behavior.* 1975.

Beebe, William. *Book of Naturalists.* 1928.

Bell, W. G. *The Great Plague of London in 1665.* 1924.

Boccaccio, Giovanni, *The Decameron.* 1982. As translated by Mark Musa and Peter Bondanella.

Bowsky, William. *The Black Death; A Turning Point in History?* 1971.

Brinton, Crane. *Portable Age of Reason Reader.* 1965.

Burnet, F. Macfarlane, and David O. White. *A Natural History of Infectious Disease.* 1972.

Burton, Maurice. *Animal Legends.* 1957.

Calhoun, J. B. *The Ecology and Sociology of the Norway Rat.* Public Health Service. 1962.

Camus, Albert. *The Plague.* 1969.

Coulton, G. G. *The Black Death.* 1929.

Crawford, Raymond. *Plague and Pestilence in Literature and Art.* 1914.

Creighton, C. *A History of Epidemics in Britain.* 1891–1894.

Crowcraft, Peter C. *Mice All Over.* 1947.

Deaux, George. *The Black Death.* 1969.

Defoe, Daniel. *A Journal of the Plague Year.* 1721.

Drummond, D. *Biology and Control of Domestic Rodents.* World Health Organization. 1972.

Ellerman, J., R. W. Hayman and G. W. C. Holt. *The Families and Genera of Living Rodents.* 1941.

Elton, C. *Voles, Mice and Lemmings.* 1942.

Evans, E. R. *The Criminal Prosecution and Capital Punishment of Animals.* 1906.

Farb, Peter and George Asmelagos. *Consuming Passions: The Anthropology of Eating.* 1976.

Finch, Christopher. *The Art of Walt Disney: From Mickey Mouse to The Magic Kingdom.* 1978.

Forbush, Edward Howe. *Rats and Rat Riddance.* 1916.

Ford, Barbara. *Future Food.* 1977.

Fuller, John C. *Fever.* 1974.

Garrison, W. B. *Codfish, Cats and Civilization.* 1963.

Gasquet, F. A. *The Black Death of 1348 and 1349.* 1908.

Gregg, Charles T. *Plague!* 1978.

Haggard, Howard M. *Devils, Drugs and Doctors.* 1929.

Henschen, Folke. *The History of Diseases.* 1966.

Hinton, M. A. C. *Rats and Mice as Enemies of Mankind.* 1918.

Hirshleifer, Jack. *Disaster and Recovery.* 1966.

Hirst, L. F. *The Conquest of Plague.* 1953.

Hogarth, Alfred Moore. *The Rat: A World Menace.* 1929.

Hubbert, W., W. F. McCulloch and P. R. Schnurrenberger. *Diseases Transmitted from Animals to Man.* 1974.

Hudson, W. H. *Book of a Naturalist.* 1930.

Lehane, B. *The Compleat Flea.* 1969.

Link, V. B. *A History of Plague in The United States of America.* Public Health Service. 1955.

Lorenz, Konrad. *On Aggression.* 1963.

Major, R. H. *Disease and Destiny.* 1939.

McNeill, William H. *Plague and Peoples.* 1976.

Munn, N. L. *Handbook of Psychological Research on the Rat.* 1950.

Nicholes, Laurens J. *Vandals of the Night.* 1948.

Nohl, Johannes. *The Black Death.* 1961.

Pollitzer, Robert. *Plague.* World Health Organization. 1954.

Pratt, H. D., and W. H. Johnson. *Sanitation in the Control of Insects and Rodents.* U.S. Public Health Service. 1975.

Pratt, H. D., B. F. Bjornson, and K. S. Littig. *Control of Domestic Rats and Mice.* U.S. Public Health Service. 1979.

Prinzing, F. *Epidemics Resulting from Wars.* 1916.

Ratcliff, J. D. *Modern Miracle Men.* 1960.

Rodenwaldt, Ernst. *World Atlas of Epidemic Diseases.* 1952–56.

Rodwell, James. *The Rat: Its History and Destructive Character.* 1858.

Sandersen, I. T. *Animal Tales.* 1945.

Schickel, Richard. *The Disney Version.* 1978.

Shrewsbury, J. F. D. *A History of Bubonic Plague in the British Isles.* 1970.

Siegfried, André. *Germs and Ideas.* 1965.

Skinner, B. F. *The Behavior of Organisms.* 1938.

Skinner, B. G. *Beyond Freedom and Dignity*. 1971.
Stilley, Frank. *The $100,000 Rat*. 1975.
U.S. Center For Disease Control. *Focus Rodent Control*. 1978.
U. S. Communicable Disease Center. *Rat Borne Diseases*. 1949.
Ziegler, Philip. *The Black Death*. 1969.
Zinnser, Hans. *Rats, Lice and History*. 1935.

Many other books and stories, factual and fictional, are noted in the text. By rough calculation anyone attempting to read *all* the literature available on rats would have to read two or three articles a day for the whole of an average life span. Of course I have compromised, but besides interviewing experts on rodents and rat control, I consulted thousands of newspaper stories, magazine articles, and scientific studies, too numerous to list here. For just one example, the complete back issues of three newspapers dating back several centuries were examined. Particularly helpful for recent rat-control progress around the world was the masterly article "The Rat, Lapdog of the Devil" by Thomas Y. Canby (*National Geographic*, July 1977), which also has the best illustrations I've seen in one place on the subject. For specialized information on rat control, one should consult the studies mentioned in Pratt et al., *Control of Domestic Rats and Mice*, listed in the bibliography. Scientific studies on the rat are covered by the hundreds of thousands in indexes that can be found in any reference library, but a good start for the beginner might be Barnett, *The Rat*, listed in the bibliography, which alone lists some 738 studies.

# INDEX

# INDEX

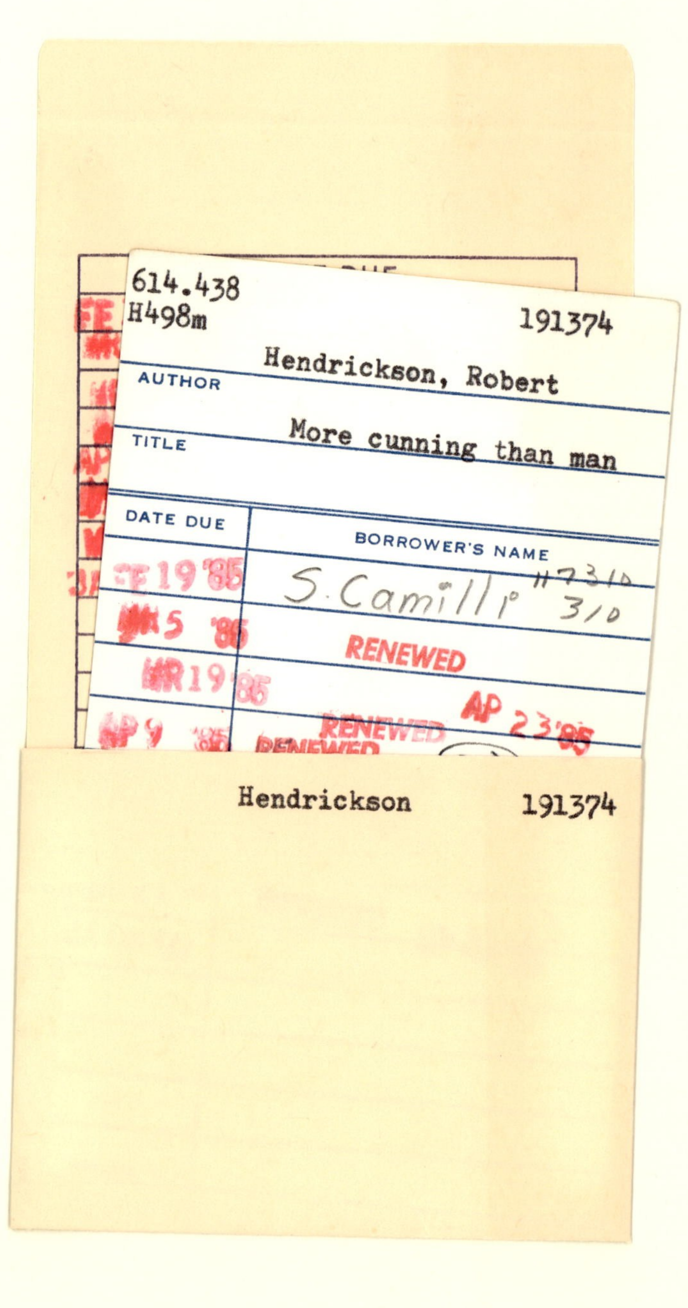
614.438
H498m
191374
Hendrickson, Robert
AUTHOR
More cunning than man
TITLE
DATE DUE
BORROWER'S NAME
RENEWED
Hendrickson
191374